Judith McKell

Problem Solving in Biology

Problem Solving in
BIOLOGY

Team Co-ordinator

James Torrance

Writing Team

James Torrance
James Fullarton
Clare Marsh
James Simms
Caroline Stevenson

Diagrams by James Torrance

Hodder & Stoughton

LONDON SYDNEY AUCKLAND TORONTO

ISBN 0 340 50843 4

First published 1989
Fourth impression 1991

Typeset by Tradespools Ltd, Frome, Somerset.
Printed in Great Britain for Hodder and Stoughton
Educational, a division of Hodder and Stoughton Ltd,
Mill Road, Dunton Green, Sevenoaks, Kent by
Thomson Litho Ltd, East Kilbride, Scotland.

Contents

Preface

This book consists of a comprehensive bank of exercises intended for use by pupils studying Standard Grade and GCSE Level Biology courses and required therefore to develop skills in problem solving.

It is the aim of the book to give pupils practice in overtaking the extended grade criteria: handling and processing information; evaluating procedure and information; and drawing conclusions and making predictions.

The material is differentiated with some problems pitched at General Level, some carrying extra, more difficult questions and some aimed solely at the most able pupils.

The items are arranged into 25 Chapters. This allows for maximum flexibility of use by providing ready-made end-of-topic tests, extensions to classwork and homework assignments.

Section 1 The Biosphere

1 Investigating an ecosystem

1 The following graph shows the range of pH within which each of six soil animals is found to occur.

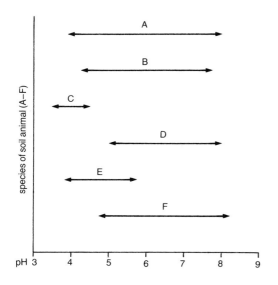

a) Which species occurs over the widest range of pH conditions?
b) Which species appears to be least tolerant of acidic conditions?
c) Which species is most tolerant of alkaline conditions?
d) Which species can survive in conditions of pH below 3.75?
e) How many species can tolerate pH 4?
f) Which of the following pH values can be tolerated by (i) the largest number (ii) the smallest number of species?

       3.5   4.5   5.5   6.5

2 The following table shows the number of different species of flowering plant and snail present in three areas of North America.

area	latitude		flowering plants	snails
1	50°	decreasing	650	30
2	40°	distance from	1625	91
3	30°	equator	2100	172

a) what effect does latitude have on the number of different species of flowering plant and snail present?
b) Suggest a reason for this trend.

3 The following diagram shows a piece of ground viewed from above. It was sampled by taking four line transects (AB, CD, EF and GH) and recording the type of plant present at regular intervals along each transect.

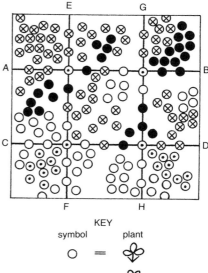

1

a) Which of the four transects corresponds to the following diagram?

b) From the results of this one transect alone, which plant type seems to be most abundant? Give its symbol.

c) Look again at the diagram of the piece of ground. Is the plant that you gave as your answer to (b) in fact the most abundant?

d) Describe a more accurate method of estimating which plant type is the most abundant in the area.

quadrats in randomly chosen sites
(each quadrat encloses one square metre)

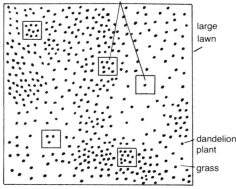

large lawn

dandelion plant

grass

4 The lawn in the above diagram measures 10 metres in length by 10 metres in breadth.

a) Calculate the area of the lawn.

b) Estimate the total number of dandelion plants growing on the lawn using only the information provided by the randomly chosen quadrats.

The owner decided to try to remove the dandelions by spraying the lawn with selective weedkiller. The following diagram shows the lawn several weeks after spraying.

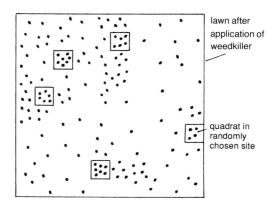

lawn after application of weedkiller

quadrat in randomly chosen site

c) Using the same method, estimate the total number of dandelion plants now growing on the lawn.

d) From a consideration of your results so far, does the weedkiller appear to have been effective? Explain your answer.

Taking an overall view of the lawn before and after spraying, the weedkiller does appear to have been at least partly successful.

e) (i) State the source of error in the sampling technique that was responsible for failing to show up this difference.

Extra Question

(ii) Explain how this error could be reduced to a minimum in a future investigation using the same sampling technique.

5 The graph below shows the effect of speed of water movement on the numbers of two species of insect larvae, A and B, found at the surface of a river.

a) At which range of water speed was the greatest number of species A recorded?

b) From which range of water speed was species A absent?

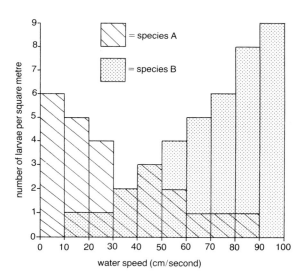

c) At which ranges of water speed were equal numbers of species A and B found?

d) For each species, state the relationship that exists between numbers present and the given ranges of water speed.

Extra Question

The following table shows the sizes of inanimate objects moved by different water speeds.

water speed (cm/second)	diameter of object (mm)	example of object
10	0.2	mud particle
25	1.3	sand particle
50	5.0	gravel particle
75	11.0	coarse gravel
100	20.0	pebble

e) Which species of insect larva would you expect to be more numerous in a stretch of river where the water speed is able to move particles of coarse gravel?

---

6 Two boys were asked to investigate the effect of light intensity on the distribution of meadow buttercup plants at the edge of the oak wood shown in the diagram below.

They pegged out a string line from point X outside the wood to point Y inside the wood. Along this line transect at points 1–10 they placed a metre square quadrat and counted the number of meadow buttercup plants present in each quadrat. This number was given an abundance score as shown in the accompanying table.

number of buttercup plants	abundance score	symbol
26 or more	abundant	◖
11–25	frequent	◖
6–10	occasional	◖
1–5	rare	◔
0	absent	◯

The boys also measured the light intensity falling on each quadrat by taking one reading using a light meter which gave readings on an 8 point scale of A–H where A = dimmest and H = brightest.

quadrat	1	2	3	4	5	6	7	8	9	10
light intensity	H	H	H	H	F	E	D	C	C	C
abundance of meadow buttercups	◖	◖	◖	◖	◖	◖	◖	◔	◯	◯

a) What abiotic factor was measured in the investigation?
b) In what way does the abiotic factor gradually change along the line transect from quadrat 1 to 10?
c) In what way does the abundance of buttercup plants change along the line transect from quadrat 1 to 10?
d) What relationship appears therefore to exist along the transect between abundance of buttercups and light intensity?
e) Suggest a possible reason for this apparent relationship.

Extra Questions

f) Considering that some parts of the ground at the edge of a wood on a sunny day are lit up by patches of bright sunlight whereas others are in the shade, spot a shortcoming in the technique used by the boys to measure light intensity.
g) Suggest how the shortcoming that you identifed in question (f) could be overcome.
h) It is possible that some other abiotic factor is wholly or partly responsible for the distribution of the buttercup plants. Suggest TWO such factors.

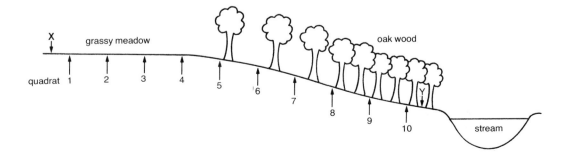

2 How it works

1 Four types of plant typical of a woodland ecosystem are:

(i) trees that produce flowers and fruit (nuts) and shed their leaves in autumn;
(ii) mosses that have leaves but no flowers and grow at the bases of the trees;
(iii) fungi that lack green leaves and feed on dead leaves from the trees;
(iv) small plants that grow under the trees and make their flowers in spring before the conditions become too shady.

a) Match each of these descriptions with the four plants given in the following chart.

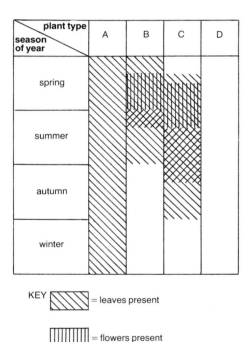

KEY

= leaves present

= flowers present

= seed and fruit present

b) Which of these plants could provide a habitat and a food supply for a population of squirrels?

2 A one-year-old carp fish has a mass of 30 g on average. A loch was stocked with a hundred of these young fish.

a) What was the total mass of this population when first put in the loch?
b) If a carp increases in mass by 50% each year, what was the total mass of the 100 fish after one year (assuming that they all survived)?
c) At the end of the first year 60% of the carp were caught by fishermen. What was the total mass of carp remaining in the loch?

3 The accompanying diagram shows the number of units of energy (in kilojoules/m²/year) that are transferred from organism to organism in a pond food chain.

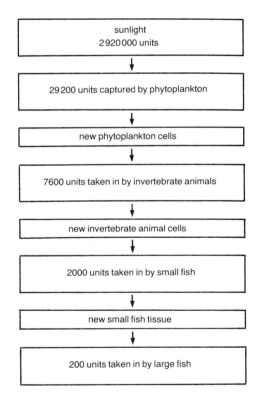

a) What percentage of the sunlight energy is successfully captured by the phytoplankton?
b) What percentage of energy is lost between intake of energy by small fish and intake of energy by large fish?

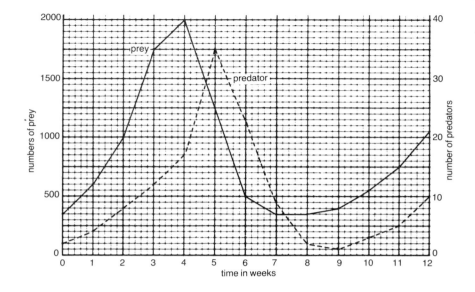

4 The graph above shows the results from a study of the populations of two organisms, a predator and its well-fed prey, over a period of several weeks.
a) How many weeks did the population of prey organisms take to reach its maximum size?
b) How many weeks did the population of predators take to go from maximum numbers to minimum?

Extra Questions

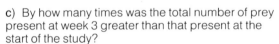

c) By how many times was the total number of prey present at week 3 greater than that present at the start of the study?
d) What was the total number of predators present at week 6?
e) Account for (i) the decrease in prey and (ii) the increase in predators during the period between week 4 and week 5.

5 The following diagrams represent the population distributions for a developed country and a developing country.

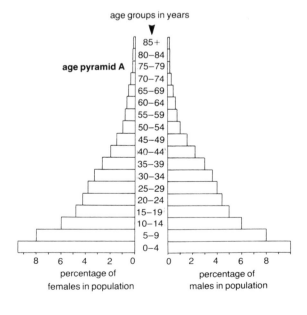

age groups in years

age pyramid A

85+	
80–84	
75–79	
70–74	
65–69	
60–64	
55–59	
50–54	
45–49	
40–44	
35–39	
30–34	
25–29	
20–24	
15–19	
10–14	
5–9	
0–4	

8 6 4 2 0   0 2 4 6 8
percentage of   percentage of
females in population   males in population

age groups in years

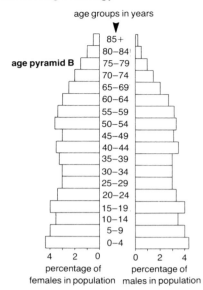

age pyramid B

4   2   0   0   2   4
percentage of   percentage of
females in population   males in population

a) In age pyramid A, (i) what percentage of the population are female aged between 5 and 9 years? (ii) which age group makes up exactly 6% of the male population?
b) In age pyramid B, (i) what percentage of the female population are aged 65 and over? (ii) which sex has the larger percentage reaching old age?

Extra Question

c) Which age pyramid represents the population distribution of the developing country? Give TWO reasons for your answer.

6 The accompanying table refers to three different countries of the world.
   a) Copy and complete the table.

	Country		
	X	Y	Z
% birth rate	2	4	5
% death rate	1	2	
% overall population growth	1		2.5

Extra Question

b) Which of these countries is likely to have both the most effective birth control measures and the best medical services? Explain your answer.

7 The table below gives the results from a plant competition experiment where five groups of pea plants (each differing in number from the others) were grown in areas of similar fertile soil measuring ¼ square metre.

number of plants per ¼ square metre	average number of pods per plant	average number of seeds per pod
20	8.3	6.0
40	6.8	5.9
60	3.9	6.2
80	2.7	5.9
100	2.1	6.0

a) Plot the data in the table as two line graphs using graph paper similar to that in the diagram which shows how to lay out the axes.

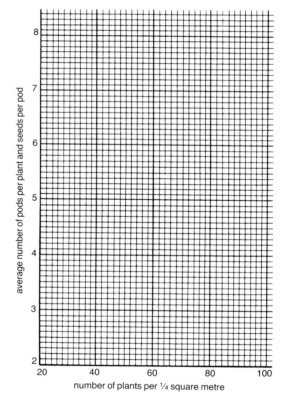

b) Which feature appears to be unaffected by competition between neighbouring pea plants?
c) (i) Which feature appears to be affected by competition?
(ii) In what way is this feature affected as density of plants in a ¼ square metre increases?
(iii) Suggest TWO factors that neighbouring pea plants may be competing for.

Extra Questions

d) Calculate the total number of seeds produced by
   (i) 20 plants on a ¼ square metre;
   (ii) 100 plants on a ¼ square metre.

e) Since seed mass is found to be unaffected by competition, which number of plants, 20 or 100, would be the better number to grow per ¼ square metre? Explain your answer.

f) It is possible that these results are unusual and not typical of pea plants in general. What should now be done to check the validity of these results?

8 The following graph shows the distribution of three species of organism at different depths in a loch where they form a food chain.

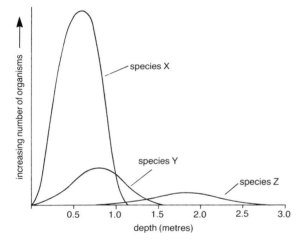

a) Which species has the widest vertical distribution in the loch?

b) Which species is the producer? Give TWO reasons to support your answer.

c) Identify the primary and the secondary consumers. Explain your choice in each case.

9 Imagine a sailor shipwrecked on a barren rocky island lacking top soil. From the ship's cargo he has managed to salvage one live hen and a bag of wheat grains.

a) To make these, his only resources, last for as long as possible, which of the following courses of action should he take?

   A eat the hen and then eat the wheat
   B feed all of the wheat to the hen, eat its eggs and then eat the hen when the wheat runs out
   C share the wheat with the hen and then eat the hen when the wheat is finished

b) Justify your choice of answer.

3 Control and management

1 Repeated sampling of the water in the river shown in the diagram was done at sample points 1–5 during the years 1980 and 1984.

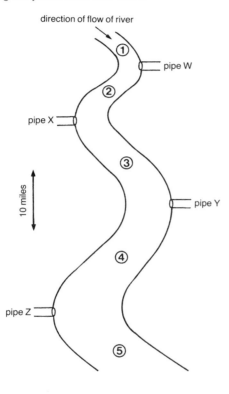

direction of flow of river

pipe W

pipe X

pipe Y

pipe Z

10 miles

The results are summarised in the following table where:

█ = many ▒ = few ☐ = none

a) (i) From which pipe was untreated sewage from an overloaded sewage works being discharged into the river in 1980? Explain your choice of answer.
(ii) Suggest a possible change that could have occurred to this sewage works during 1981–83 to account for the results obtained in 1984.
b) In 1984 a paper mill began discharging untreated organic waste into the river through pipe Z.
Describe the effect this had on
(i) the type of animal most commonly found at sample site 5;
(ii) the oxygen concentration of the water at sample site 5.

2 The accompanying table shows the amount of sulphur dioxide produced by human activities and released into the atmosphere by an industrial European country over a period of sixty years.
a) Calculate the average amount of sulphur dioxide produced per year during the period shown.
b) Present the information as a line graph.
c) Draw TWO conclusions from the data.

year	amount of sulphur dioxide (million tonnes per year)
1920	3.8
1930	3.6
1940	4.4
1950	5.1
1960	6.3
1970	6.8
1980	5.0

		1980					1984				
sampling point → indicator species ↓		1	2	3	4	5	1	2	3	4	5
stonefly nymph		█				█	█	█			
caddis fly larva		▒			█	▒	▒	▒	▒		
bloodworm		▒		█	▒	▒	▒	▒	▒		█
rat-tailed maggot		▒	█	▒	▒		▒	▒	▒	█	

3 Read the passage and answer the questions that are based on it.

In the USA and many other countries, there is an enormous demand for cheap fastfoods such as beefburgers. The price of North American home-grown beef has rocketed in recent years, making it too expensive for use in low price burgers. Instead, cheap beef is imported from Central American countries such as Mexico. This beef comes from cattle raised on grasslands that used to be tropical forests.

It is to these same Central American forests (the ones that still remain) that North American songbirds such as the warbler migrate in autumn. However, as the forests are cleared, more and more songbirds fail to find a suitable overwinter habitat and therefore perish. Fewer songbirds return the next spring to North America where their food, countless species of young insect, are poised ready to devour crop plants. Until now predation by songbirds has kept the number of pests at a tolerable level but what of the future with songbird numbers declining by up to 4% per annum?

If Mexican ranchers could be persuaded to make better use of existing pasture lands, they would be able to produce about twice as much beef without clearing any more forests. But it is difficult to persuade landowners to invest money in fertilising and improving existing tired grasslands when new, temporarily rich, pastures can be quickly developed by clearing forested areas. However, in the long run this may prove to be a more realistic approach than attempting to persuade the average American fast food devotee to turn vegetarian.

a) Why is Central American beef used for burgers instead of home-grown North American meat?
b) Construct a food chain from the information given in the first paragraph.
c) 'Removing one link in a food chain can lead to disaster'. Discuss the truth of this statement using an example contained in the passage.
d) (i) Give TWO possible solutions to the problem mentioned in paragraph 3.
(ii) Which ONE of these does the author suggest is less likely to be successful? Do you agree? Explain why.

4 The following information refers to a polluted river. Two readings have been omitted from the table.

river organism	condition of river water				
	very clean	clean	fairly clean	dirty	very dirty
green algae	1	2	3	4	4
trout	3	1	0		0
waterweeds	1	3		3	1

KEY TO ABUNDANCE LEVELS	
point on scale	description of population
0	absent
1	scarce
2	moderate
3	plentiful
4	abundant

a) Describe the relationship that exists between the number of green algae and the condition of the river water.
b) Give the word that describes (i) the population of trout in clean water, (ii) the population of waterweeds in dirty water.
c) From the choice given in brackets, select the appropriate number to indicate the most likely abundance of (i) trout in dirty water (0, 1, 2, 3), (ii) waterweeds in fairly clean water (0, 1, 2, 3).

5 Read the passage and answer the questions that are based on it.

Crop yield by a piece of land is directly related to soil depth. Deeper soils produce better crops because they contain more water and mineral salts necessary for healthy plant growth.

Soil is a natural resource which renews itself only very slowly at a rate of 0.5 tonnes per hectare per year. However, annual soil loss in some parts of Britain is occurring at a rate of 30 tonnes per hectare per year. Soil that used to be 100 cm deep is now reduced to a thin layer of only 20 cm. This problem has become especially acute in recent years owing to farmers planting winter cereals which leave large areas of ground bare and susceptible to soil erosion during the winter months. This is especially true of steeply sloping land which has been ploughed in a downslope line rather than along the contours (see diagram).

land ploughed in vertical downslope lines

land ploughed in horizontal lines along contours of slope

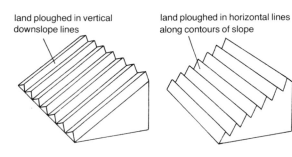

Instead of being offered financial incentives to conserve soil, farmers are encouraged by generous government subsidies to grow these high-yielding autumn-planted winter cereals on any available land. This results in overproduction of grain which then costs millions of pounds to store as a grain 'mountain' while the land that used to be over-wintered as grass continues to lose its top soil to wind and rain.

a) What natural resource is referred to in the above passage?
b) Explain why crop yield is directly related to soil depth.
c) By how many times is soil being lost at a faster rate than it is renewing itself in those parts of Britain referred to in paragraph 2?
d) (i) What name is given to this process of soil loss?
(ii) Name TWO environmental factors that bring about this soil loss.
e) Give an example of a method of soil preparation carried out by some farmers that makes the problem of soil loss even worse.
f) Why do farmers plant winter cereals on every available hectare of soil?
g) Suggest a possible remedy to the problem of soil loss at (i) a local level, (ii) a national level.

6 The map below shows a region of coastline close to where a giant oil tanker was wrecked at sea. Prior to the disaster, the shallow waters of the coastline provided a rich source of edible crabs. Oil does not kill the crabs but harms their flesh, making them unsaleable. The extent of the shaded sector at each sample site represents the proportion of crabs with diseased flesh after the disaster.

a) Which sample site had the highest number of crabs?
b) In which sample site were the crabs only rarely found?
c) Describe the abundance level of crabs at sample site T.
d) Name the agent of pollution that affected the crabs.
e) In which sample site were fewest crabs affected by the pollutant?

Extra Questions

f) Give two possible reasons for your answer to question (e).
g) In which sample site were most crabs affected relative to the population size present? Suggest why.

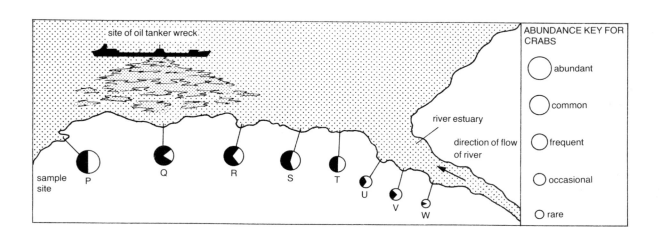

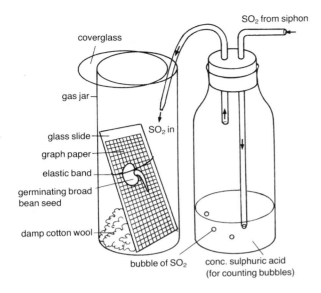

coverglass

gas jar

glass slide

graph paper

elastic band

germinating broad
bean seed

damp cotton wool

SO₂ from siphon

SO₂ in

bubble of SO₂

conc. sulphuric acid
(for counting bubbles)

The increase in length of each young root was measured after five days and the results recorded as shown in the following table

number of SO₂ bubbles	increase in length of root after 5 days (mm)
5	6
10	2
20	0.5

a) Suggest the reason for using (i) the graph paper, (ii) the damp cotton wool.
b) What control should have been included in the experiment?
c) Predict the effect of using 30 bubbles of SO₂.

Extra Questions

d) Identify a possible source of error that could occur when setting up this experiment.
e) Apart from attempting to overcome this source of error and including a control, suggest a further improvement that could be made to the experiment.

7 The above apparatus shows a method used to investigate the effect of sulphur dioxide (SO₂) gas on the growth of young roots.

Three gas jars, each containing a germinating broad bean seed, were set up as shown. Using a SO₂ siphon, 5 bubbles of SO₂ were passed into the first gas jar. The second gas jar received 10 and the third jar 20 bubbles of SO₂. Each gas jar was sealed using a coverglass.

4 Investigating living cells

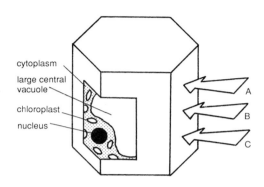

cytoplasm

large central
vacuole

chloroplast

nucleus

A

B

C

1 The diagram above shows a 3-dimensional view of
a plant cell with a 'window' cut out to reveal some of
its internal structures.

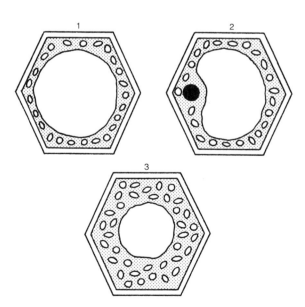

1

2

3

Imagine that the cell has been sliced across at
planes A, B and C. Match these with the above
diagrams of the cell's internal structure.

2 The unit used to measure cell size is the
micrometre. 1 millimetre = 1000 micrometres.
a) A human red blood cell is 7 micrometres in
diameter. Express this as a decimal fraction of a
millimetre.
b) If a cell is 0.008 millimetres long, what is its
length in micrometres?

3 Present the information given in the following table
as a bar chart.

cell type	cell length or diameter (in micrometres)
onion epidermis	150
human egg	100
sea urchin egg	70
Paramecium	50
human liver	20
yeast	8
human red blood corpuscle	7
Bacillus bacterium	3

4 Two purposes of brown iodine solution in a
biological laboratory are: to stain structures (if a
structure absorbs more iodine solution than its
surroundings, it becomes a darker brown colour);
and to test for starch (if starch is present, it turns
blue-black in the presence of iodine solution).
The diagram opposite shows three types of plant
cell before and after the addition of iodine solution.

a) (i) Which cellular structure became stained
brown?
(ii) In which cell types was the process of staining
observed in this experiment?
b) (i) Which cell type contains starch grains?
(ii) Give a reason for your answer.

	before	after
green cell from *Elodea* leaf		chloroplast, cell wall, nucleus (dark brown)
cell from potato tuber		blue-black structure, cytoplasm, cell wall
cell from onion epidermis		cell wall, cytoplasm, nucleus (dark brown)

presents a large surface area through which oxygen can pass into the cell before being transported to all parts of the body. Thanks to its shape, a red blood cell is very good at this job.

A group of similar cells working together and carrying out a particular function is called a tissue. For example, muscle tissue brings about movement and nerve tissue carries messages. A group of different tissues in turn make up an organ. For example, the heart is made of different tissues such as muscle, connective and nerve tissues working together to pump blood around the human body.

a) Explain why a cell is often described as 'the basic unit of life'.
b) Lines 6–7 refer to '. . . all the characteristics of living things . . .'. List as many of these as you can.
c) Distinguish between unicellular and multicellular living things.

5 Read the passage and answer the questions which are based on it.

If the nucleus of a cell, or some other structure such as a chloroplast, is removed from the cell, the isolated part cannot survive on its own. A complete cell is the smallest unit that can lead an independent life. This is neatly illustrated by unicellular *Amoeba*, a tiny animal which shows all the characteristics of living things despite the fact that it consists of only one cell.

Multicellular animals and plants are made of more than one cell. A human adult's body is composed of approximately 60 billion cells. Instead of each one of these cells performing every function vital for life, it is more efficient for certain cells to become specialised to do particular jobs.

A red blood cell, for example, is shaped like a biconcave disc as shown in the diagram. This shape

Extra Questions

d) Another way of expressing the number 1000 is 10^3.
(i) A human baby is made up of about 2 billion cells. Express this number in the same way (where 1 billion = 1 million million).
(ii) By how many times is the number of cells in a human adult greater than that in a baby?
e) Why is a red blood cell a good example of a cell whose structure is ideally suited to the job that it does?
f) What general name is given to a group of similar cells working together to do a particular job?
g) Human skin is composed of many structures such as blood vessels, epidermal cells, nerve endings, sweat glands, etc. Is skin a tissue or an organ?
h) Arrange the following in order of increasing complexity: organ, cell, human body, tissue.

6 The average length of a certain type of cell was found to be 150 micrometres. How many of this type of cell would need to be laid end to end to form a line the same length as this page? (Answer to nearest whole cell.)

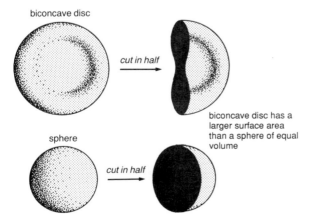

biconcave disc

cut in half

biconcave disc has a larger surface area than a sphere of equal volume

sphere

cut in half

5 Investigating diffusion

1 The diagram shows an experiment set up to investigate the process of diffusion. After ten minutes a distinct blue-black colour was found to be present inside the visking tubing sausage but not outside it.

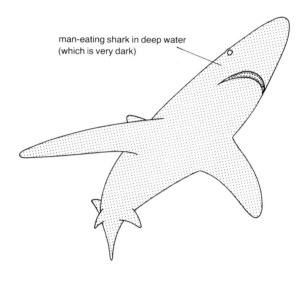

man-eating shark in deep water (which is very dark)

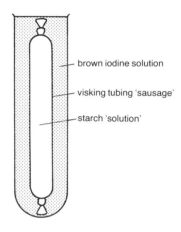

- brown iodine solution
- visking tubing 'sausage'
- starch 'solution'

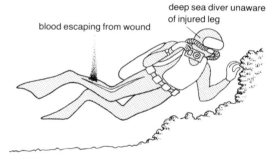

deep sea diver unaware of injured leg

blood escaping from wound

a) Rewrite the following sentence choosing the correct word.
Iodine solution has reacted with starch inside/ outside the visking tubing 'sausage'.
b) It was concluded from the experiment that molecules of one of the two liquids were small enough to diffuse through the visking tubing membrane. Identify this liquid.
c) It was concluded that molecules of one of the liquids were too large to diffuse through the membrane. Identify this liquid.

Extra Questions

d) An alternative explanation of the results is that the membrane allows molecules of any size to pass in but not out.
(i) Describe the experiment that you would set up to investigate this theory.
(ii) Explain how you would know from your results whether the theory is true or false.

2 Write a short paragraph to explain why the deep sea diver shown in the diagram is in danger. Use all the following words and phrases in your answer: low concentration, particles of blood, diffusion, high concentration, deep water.

3 Five identical cylinders of fresh potato (A–E) were immersed in sugar solutions of different concentration for three hours and then reweighed. The change in mass of each cylinder was expressed as a percentage of its original mass and the results recorded in a bar chart as shown in the diagram opposite.

   Which cylinder has been in sugar solution nearest in water concentration to that of potato cells?

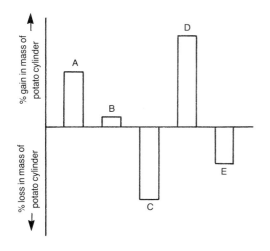

5 When sugar is sprinkled onto fresh strawberries, liquid is found to collect in the dish. Suggest a reason for this.

6 A piece of rhubarb epidermis tissue was immersed for 5 minutes in each of three liquids (A, B and C), one after the other, and the cells examined at the end of each 5 minute period. The following diagram shows the appearance of one typical cell from the tissue after each immersion.

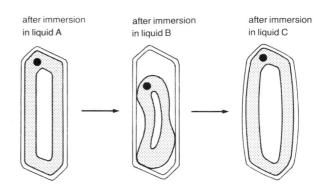

after immersion in liquid A   after immersion in liquid B   after immersion in liquid C

4 In the experiment shown below, the level of sugar solution was found to rise up the tube. The level was read at regular intervals and the results graphed as shown.

a) What evidence is there from this experiment that visking tubing is a selectively permeable membrane?

b) What was the height of the sugar level at minute 4?

c) How much longer did it take for the sugar level to reach the top of the tube?

d) Predict the effect of using a more dilute sugar solution on the time taken for the level to reach the top of the tube.

a) Match the three liquids with the following: water, dilute sugar solution, concentrated sugar solution.

b) Immersion in which liquid brought about plasmolysis?

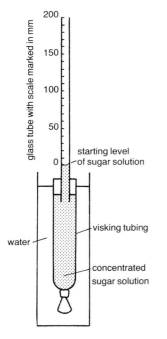

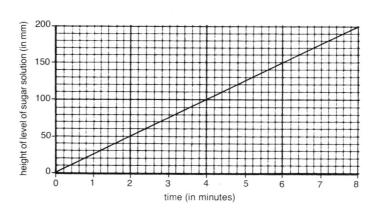

6 Investigating cell division

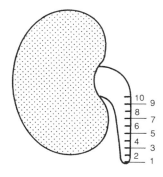

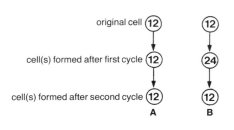

1 The above diagram shows a germinating broad bean seed. Its young root has been marked at 1mm intervals with waterproof ink.

It was allowed to grow for a further three days and was then carefully examined.

Mitosis had occurred between marks 1 and 2 only. The cells in the region marked 2–5 had not divided but instead had become elongated. The cells in the region marked 5–10 had neither divided nor elongated.

Which of the following diagrams best represents the young root three days after receiving its ink marks?

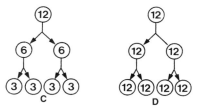

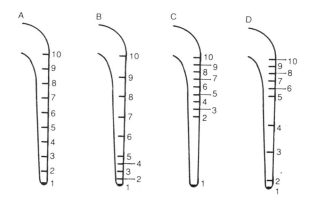

2 Each normal body cell in the kangaroo contains 12 chromosomes. Which of the following diagrams correctly represents two successive cycles of mitosis and cell division? (The numbers refer to the chromosomes present in each cell.)

Extra Questions

3 A student observed cells dividing under a microscope and recorded her results as shown in the table below.

The distance to which the table refers is the average distance between the centromeres of the chromosomes and the poles of the spindle.

time (minutes)	distance (micrometres)
0	20
5	20
10	9
15	2
20	0

a) Plot the data as a line graph.
b) During which time interval did sister chromatids begin to move apart?
c) Explain how your arrived at your answer to (b).

4 The diagram below shows some of the stages in the development of a certain multicellular animal. A series of cell divisions of the fertilised egg in a vertical and horizontal plane results in the formation of an embryo. Copy and complete the diagram below.

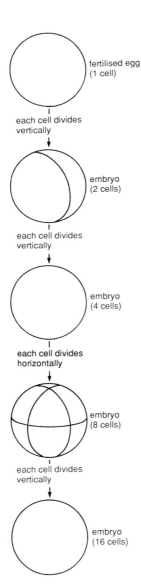

fertilised egg
(1 cell)

each cell divides
vertically

embryo
(2 cells)

each cell divides
vertically

embryo
(4 cells)

each cell divides
horizontally

embryo
(8 cells)

each cell divides
vertically

embryo
(16 cells)

7 Investigating enzymes

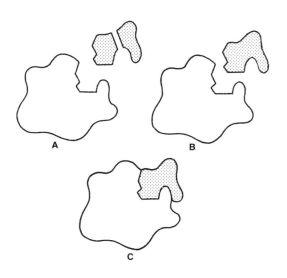

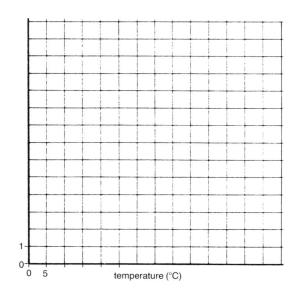

1 The above diagrams show the stages that occur during the action of an enzyme which promotes a breakdown reaction. Using the three letters given and arrows, draw a flow diagram to indicate the correct sequence in which the three stages would occur.

2 The following table gives the results from an experiment involving the breakdown of a food by an enzyme.

temperature (°C)	mass of food broken down (mg/hour)
5	3
15	9
25	17
35	21
45	18
55	1

a) Using similar graph paper and the axes shown in the diagram (which have been only partly completed), present the results as a graph by plotting the points and joining them into a curve.
b) State the temperature at which the enzyme is most active.
c) In your opinion, what does the graph suggest about the amount of food that would be broken down at 65 °C?

3 The experiment in the diagram below was set up to investigate the effect of the enzyme salivary amylase on starch 'solution'. Two equal lengths of visking tubing were filled with 1% starch 'solution', rinsed and placed in test tubes of water. Both test tubes were kept in the water bath at 37 °C for one hour.

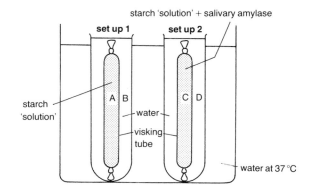

Samples of liquid from regions A, B, C and D were tested for sugar and starch at the start and after one hour. The results are shown in the following table. (+ = food present, − = food absent)

	sugar test				starch test			
	A	B	C	D	A	B	C	D
at start	−	−	−	−	+	−	+	−
after 1 hour	−	−	+	+	+	−	−	−

a) Which set-up is the control?
b) Name THREE features of the experimental procedure which are essential for a valid comparison of set-ups 1 and 2 to be made.

Extra Question

c) Draw TWO conclusions from the results of the experiment.

4 Plant amylase is an enzyme which digests starch to simple reducing sugar. In an experiment, an equal volume of plant amylase was added to a hole in the centre of each of six petri dishes containing starch agar. Each plate was kept at a different temperature for 24 hours and then the surface of starch agar flooded with iodine solution. The diameter of the non blue-black zone round each hole was measured as an indication of the amount of enzyme activity (since the starch in this zone had been digested).
    The following diagram shows the results plotted on axes as 6 points.

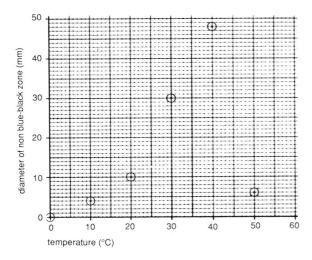

temperature (°C)

a) Using similar graph paper, copy and complete the graph by joining the points with a curve (NOT straight lines).
b) Copy and complete the following table to show the data which gave rise to the 6 plotted points.

	diameter of non blue-black zone (mm)
0	
10	
	10
30	
	48
50	

c) Using your graph, state the diameter of the non blue-black zone that would have resulted at (i) 25°C and (ii) 35°C.
d) Predict the diameter of the non blue-black zone that would have been formed in a plate kept at 60°C. Explain your answer.

5 1 g of roughly chopped liver was added to hydrogen peroxide solution at different pH values and the time taken to collect 1 cm^3 of oxygen was noted in each case. The results are given in the following table.

pH of hydrogen peroxide solution	time to collect 1 cm^3 of oxygen (seconds)
6	105
7	78
8	57
9	45
10	52
11	66
12	99

a) Present the results in the form of a line graph with pH on the horizontal axis.
b) From your graph state the pH at which the enzyme was (i) most active, (ii) least active.

Extra Questions

c) Of the pH values used in this experiment, which is the optimum for the enzyme present in the liver cells?
d) How could you obtain an even more accurate measurement of the optimum pH at which this enzyme works?

8 Investigating aerobic respiration

1 The experiment shown in the accompanying diagram was used to compare the amount of heat energy released by two foodstuffs. State TWO factors which must be kept constant for a valid comparison to be made between the two food samples.

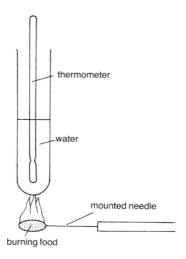

thermometer

water

mounted needle

burning food

2 The following results were obtained using a food calorimeter.
(4.2 kJ = amount of energy required to raise the temperature of 1000 g water by 1 °C)

foodstuff	state	number of kJ released on burning 1 g of food
fish	fried	8.4
egg	fresh	6.8
kidney	fresh	4.2
peas	tinned	2.7
orange	fresh	1.5

a) Present the information in the table as a bar chart.
b) Which foodstuff was poorest at heating the water?
c) How many grams of kidney would have to be burned to raise the temperature of 1000 g of water by 1 °C?
d) Burning 1 g of fried fish would raise the temperature of 1000 g of water by how many °C?

3 The apparatus shown in the diagram can be used to discover if soaked peas give off carbon dioxide.

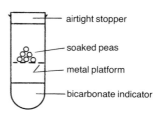

airtight stopper

soaked peas

metal platform

bicarbonate indicator

Which of the tubes shown below would be the most suitable control for this experiment?

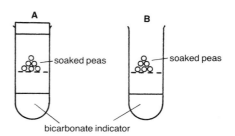

A — soaked peas

B — soaked peas

bicarbonate indicator

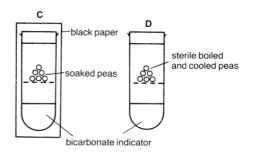

C — black paper — soaked peas

D — sterile boiled and cooled peas

bicarbonate indicator

4 Read the passage and answer the questions which are based on it.

Fresh fruits are made of living respiring cells. Once a fruit has been picked, its future life is of limited length. It soon becomes overripe and loses quality and flavour. The ever increasing demand for fresh fruit out of season presents scientists with a considerable challenge – how to keep the fruit fresh.

It has been known for many years that the storage life of apples can be extended for months by keeping them at 3 °C in an atmosphere containing a reduced level of oxygen and an increased level of carbon dioxide. However, when the apples are taken out of this cold storage, respiration rate returns to normal and soon the fruit becomes overripe.

In recent years sealed containers have been used as a method of packaging. This does slow down the ripening process because respiration decreases as the oxygen supply inside the container becomes exhausted. However, water builds up inside the carton creating ideal conditions for fungal moulds to attack and rot the fruit. A ventilated polythene bag partly overcomes this problem by letting excess water vapour escape but oxygen gets in through the holes allowing the fruit to respire and become ripe.

Ever striving to extend the shelf life of fresh fruit, scientists have now developed sophisticated plastic films that allow a much subtler control of gaseous exchange than ventilated polythene. The new wrapping film (called modified atmosphere packaging) allows water vapour out but just enough oxygen in to maintain the minimum respiration rate needed to keep the fruit healthy. Different membranes have been developed for different types of fruit because those that respire at a faster rate need a slightly more permeable membrane to enable a little more oxygen to enter and allow a healthy but still low level of cellular respiration.

a) What is the main theme of the above passage?
b) (i) By what means can overripening of apples be delayed for several months?
(ii) Suggest why this process succeeds in delaying overripening.
c) (i) Inside a sealed container of tomatoes, which two end products of aerobic respiration will gradually build up in concentration?
(ii) If left undisturbed for several more days, what will probably happen to the state of the tomatoes? Why?
d) Give ONE advantage and ONE disadvantage of using ventilated polythene for packaging fruit.
e) Explain why modified atmosphere packaging is better than both sealed containers and ventilated polythene.
f) Sweet peppers are found to require a plastic film that is slightly more permeable than the one best suited for packaging tomatoes. Compare the normal respiration rates of these two fruit types.

5 The following apparatus was set up in an attempt to demonstrate that bacteria in garden soil produce carbon dioxide during respiration.

a) Spot the error in the experimental set-up and say how you would correct it.
b) Assume that the error has been corrected. Should the pump draw air through the apparatus from W to X or from X to W?

c) If bacteria are giving out carbon dioxide, in which flask of bicarbonate indicator solution will a colour change occur?

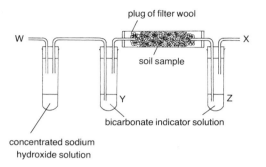

plug of filter wool

soil sample

W ——————————— X

bicarbonate indicator solution

Y   Z

concentrated sodium hydroxide solution

Extra Question ⊞

d) The experiment was run and the appropriate indicator took 1 hour to change from red to yellow. The experiment was then repeated using an equal amount of the same type of soil to which 5 cm^3 of glucose solution had been added. This time the indicator only took 30 minutes to change. Suggest why.

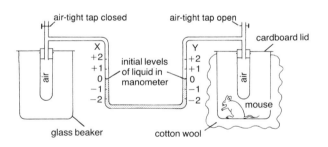

air-tight tap closed   air-tight tap open

cardboard lid

X   Y
+2   +2
+1   +1
0   0
−1   −1
−2   −2

air   initial levels of liquid in manometer   air

mouse

glass beaker   cotton wool

6 The above diagram shows an experiment using a differential air thermometer to investigate if a mouse gives out heat energy during respiration.
a) State TWO alterations that need to be made to the apparatus before starting the experiment.

Extra Question ⊞

b) Assume that the necessary alterations were made and that after 10 seconds the manometer levels were: side X = +1 and side Y = −1.

If the experiment were repeated with two similar mice in side X and one mouse in side Y, which set of results in the table below would be obtained after 10 seconds?

	Side X	Side Y
A	−1	+1
B	−2	+1
C	+1	−1
D	−2	+2

7 During an inspection, a pest control officer recorded the temperature of the grain in a grain store at different depths below the surface (without disturbing the grain). The following graph shows her results.

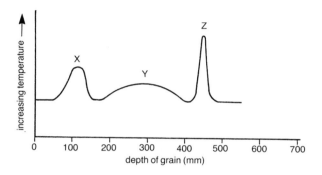

a) Pests X, Y and Z were found at the points shown. In general, what has happened to the temperature of the grain at each of these points? Explain why.
b) Which pest was found nearest to the surface of the grain?

Extra Questions

c) Which pest was most widespread in its vertical distribution in the grain?
d) Predict the form that the last part of the line graph would take if the grains at depth 550–700 mm were damp and had begun to germinate. Explain your answer.

Section 3 The World of Plants

9 Introducing plants

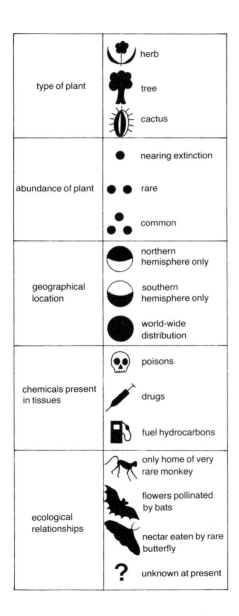

type of plant		
		herb
		tree
		cactus
abundance of plant		nearing extinction
		rare
		common
geographical location		northern hemisphere only
		southern hemisphere only
		world-wide distribution
chemicals present in tissues		poisons
		drugs
		fuel hydrocarbons
ecological relationships		only home of very rare monkey
		flowers pollinated by bats
		nectar eaten by rare butterfly
	?	unknown at present

1 The symbols given in the table are used to describe plants A–L listed below.

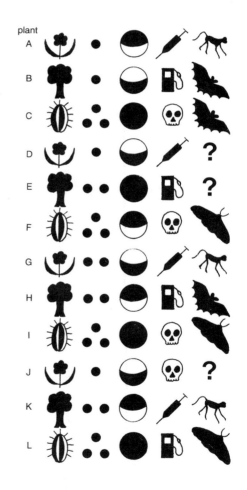

a) One of the above plants is a common, poisonous cactus with a world-wide distribution whose flowers provide food for a rare species of butterfly. Which one?

23

b) Which herb plant, found only in the southern hemisphere, is nearing extinction yet if saved may prove to be of medical use to mankind in the future?
c) Which rare plant, found only in the northern hemisphere, provides the habitat for an endangered animal species?
d) Identify the plant which contains fuel hydrocarbons yet is nearing extinction.
e) Describe plant J.
f) Name THREE features that are shared by plants A and G.
g) Compare plants C and H in detail.

plant	edible part	% protein content
broad bean	seed	7
almond	nut	20
lentil	seed	24
apricot	fruit (dried)	5
pea	seed	6

3 a) Draw a bar graph of the above information and insert the name of each edible plant part inside its own bar.
b) By how many times is the percentage of protein in almond nut greater than that present in dried apricot fruit?

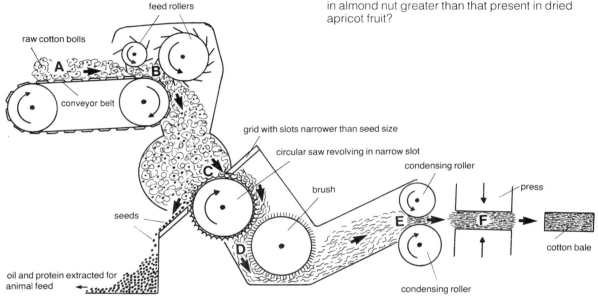

2 Cotton bolls contain tiny seeds entangled in masses of woolly-looking fibres called lint. Separation of seeds from lint is done by a machine called a gin as shown in the diagram above.
a) Which letter in the diagram indicates the point in the process where the cotton fibres are caught on teeth and pulled away from the seeds?
b) To what use are the seeds later put?
c) How many different types of rollers are named in the diagram?
d) Would the brush be turning clockwise or anti-clockwise?
e) What finally happens to the brushed cotton fibres in the above process?

4 Read the following passage.

Two parts of a plant provide food for animals: the easily-digested cell contents and the indigestible cellulose cell walls. Most herbivorous (plant-eating) mammals overcome the cell wall 'problem' by having long guts inhabited by cellulose-digesting bacteria.

The panda is a herbivore with a relatively short gut which lacks bacterial assistants capable of breaking down plant cell walls. The panda has to gain most of its food from the cell contents of the bamboo shoots and other plants that it eats. To survive it has to work continuous shifts of eight hours of feeding and four hours of sleeping, day and night.

In the cold, damp Chinese forests where the panda lives, bamboo plants produce only shoots for very long periods of time such as 60 years and then

	WEST ◄──────────────► EAST		
year	state X	state Y	state Z
1946	10.3%	9.4%	8.9%
1948	10.7	9.7	9.6
1950	11.9	11.5	10.1
1952	12.4	12.2	10.6
1954	12.9	12.6	11.2

flower, produce seeds and die. It takes up to 15 years for the new plants to develop shoots that the panda can eat.

In earlier times pandas were able to migrate from one wooded hillside that had gone to seed to another where flowering had yet to take place. However, within the last few years their ways have been barred by man clearing and cultivating the floors of the wooded valleys. Even the Wolong Panda Reserve, one of the animal's few remaining strongholds, is steadily shrinking under the ever increasing demands of local people for firewood.

a) State ONE way in which the structure of the gut of a panda differs from that of most herbivorous mammals.

b) What benefit is gained by herbivores that have bacterial 'assistants' in their gut?

c) Explain fully why a panda must spend so much of its time feeding.

d) Why do pandas suddenly have to move from one place to another after having lived there for many years?

e) What has prevented such a migration in recent years?

f) If the present situation continues, predict the fate of the panda.

5 The figures in the above table show the average percentages of protein present in one species of cereal plant growing in three different states of the USA.

Draw TWO conclusions from the information.

6 Read the following passage and study the diagrams that relate to it.

In 1987 the yeheb bush (*Cordeauxia edulis*) only grew wild in a few poor regions of Africa where many of the people were starving. Many years previously, the plant had been widespread in the Horn of Africa and its nuts had been a traditional food of the local nomadic people. In addition the plant had been an abundant source of dyes, fuel and forage for live-stock.

Of the few plants that remained in 1987, demand greatly outstripped supply; any nuts that did develop were soon eaten and the vegetation was overgrazed by goats. If this situation had been allowed to continue, the yeheb plant could have faced extinc-tion.

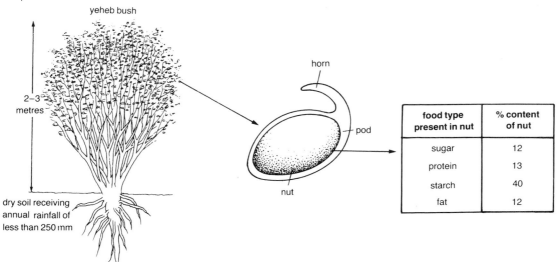

yeheb bush

2–3 metres

dry soil receiving annual rainfall of less than 250 mm

horn

pod

nut

food type present in nut	% content of nut
sugar	12
protein	13
starch	40
fat	12

AFRICA

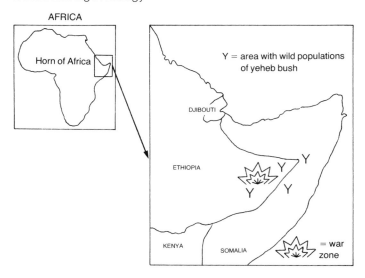

LEVEL OF ADVANCING DROUGHT IN 1987

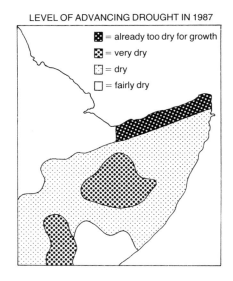

a) Name TWO countries in which wild populations of yeheb plant were still found to be growing in 1987.

b) Describe yeheb's natural habitat.

c) (i) Give FOUR traditional uses of the yeheb bush.

(ii) According to the passage, which two of these were threatening the plant's very existence in 1987?

(iii) From the maps identify two other problems that may have been causing a reduction in the number of wild yeheb bushes in 1987.

d) Construct a bar graph of the contents of a yeheb nut.

Extra Questions

e) If an average nut weighs 1.6 g, calculate the mass of starch that it contains.

f) An Ayrshire potato tuber contains on average 19% starch, 2% protein, 0.5% sugar and no fat. Would the potato be preferable to the yeheb nut as the main part of the Horn of Africa people's diet? Explain your answer.

g) Suggest why the local people did not plant the few remaining yeheb nuts and grow a crop in 1987.

h) The yeheb has now been introduced to Sudan, Yemen and Kenya. If this experiment proves to be a success, what conservation measure should be adopted when peace returns to the Horn of Africa?

7 Read the following passage.

In one particular year the world produced 300 million metric tonnes of both wheat grains and potato tubers, making them appear to be of similar importance as food crops.

However, the average water content of a potato tuber is 80% of its mass whereas that of a wheat grain is only 12%.

In addition, during the preparation of potatoes, often as much as 25% is removed during peeling and thrown away whereas the 25% of wheat grains removed during milling is used in animal feed.

a) Calculate the total weight of water present in the world crop of (i) wheat grains, (ii) potato tubers, during the year referred to in the passage.

Extra Questions

b) Give TWO reasons why, weight for weight, wheat is a more valuable crop than potato.

c) A food scientist began with 1000 g of potato tubers. He peeled them all, thus losing 25% of their mass, and then he dried the potatoes to constant weight to form potato powder. Calculate the weight of potato powder formed.

8 The graph and key below refer to the economics and management of a forest of conifers in Scotland.

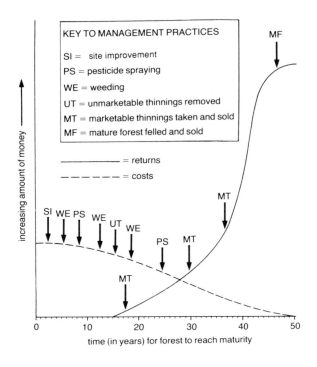

a) How many times was the crop sprayed with pesticide?
b) How many different types of management practice were applied to the forest during its first 25 years of growth?
c) In which year did costs equal returns?

Extra Questions

d) In general, what relationship exists between costs and returns over the given period of time? Explain why this is the case.
e) Suggest why weeding and thinning out unhealthy weaker trees improves the quality of the remaining trees in the long run.

10 Growing plants

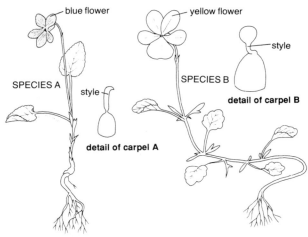

blue flower

yellow flower

SPECIES A

SPECIES B

style

style

detail of carpel B

detail of carpel A

1 a) Use the following key to identify the above two species of *Viola*.

1 style expanded into ball-like stigma 2
  style not expanded into ball-like stigma 3

2 long creeping horizontal stem present *Viola lutea*
  no long creeping horizontal stem present ... *Viola tricolor*

3 style extended into hook-like stigma 4
  style not extended into hook-like stigma .... *Viola palustris*

4 leaf stalks (petioles) hairy 5
  leaf stalks (petioles) not hairy 6

5 flowers sweetly scented .... *Viola odorata*
  flowers not sweetly scented .... *Viola hirta*

6 petals blue .... *Viola canina*
  petals white .... *Viola stagnina*

b) Describe *Viola odorata*.

2 Read the passage and answer the questions based on it.

Following pollination, the flowers of *Banksia* (a flowering plant of the Australian bush) die. However, instead of dropping off, the seeds remain firmly attached to the parent plant for years waiting for the arrival of an agent which is disastrous to most living things – a bush fire.

Whereas many eucalyptus trees survive the fire thanks to their tough insulating bark, *Banksia* seeds not only survive but actually depend on fire to open their seed coats. A seed case consists of two woody valves hinged together and it only opens when it has been thoroughly desiccated by fire.

Once the fire has passed, the paper-thin, single-winged seeds need wind to disperse them among the ashes where they germinate when the rains arrive.

a) Choose the most suitable title for the passage from the following:

A Seed dispersal in Australian flowering plants
B The role of fire in *Banksia*'s life cycle
C Flammable agents of seed dispersal in flowers
D *Banksia*'s dependence on fire for pollination

b) Choose TWO phrases from the passage that show *Banksia* to be very different from other flowering plants.

3 100 different species of flowering plants were investigated to find out which methods could be successfully used to propagate them artificially. The following results were obtained.

number of plant species	method of artificial propagation	code
25	grafting	G
5	layering	L
50	cutting	C
20	tissue culturing	T

Which of the following pie charts correctly represents this information?

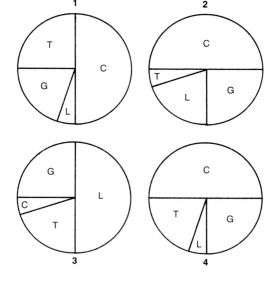

1

2

3

4

4 Read the passage and answer the questions based on it.

In a plant-cloning laboratory, all members of the workforce must wear sterile clothing. Gone are the traditional pots of soil found in greenhouse and potting shed; in their place are countless rows of gleaming culture tubes. Here thousands of copies of one plant are produced by micropropagation.

A plant with desirable characteristics (e.g. sweet fruit and disease resistance) is selected and tissue-cultured. One technique involves removing tiny pieces of bud tissue from the parent plant and growing them in sterile nutrient jelly in culture tubes. Each piece of tissue is kept in optimum conditions of light and temperature until it develops into a miniature plant. It is then checked carefully for disease before being supplied in a batch to a commercial grower who wishes to obtain a uniform crop (e.g. fruit that all ripen at the same time).

This seemingly ideal situation does, however, carry an element of risk. If a new disease-causing organism arrives that can attack the propagated plants then the whole crop may be lost.

a) Which of the following states the main theme of the passage?

A   Investigating a plant disease
B   Tissue-culturing a useful plant
C   Propagating a plant by natural means
D   Producing new varieties of a plant

b) The following flow diagram is intended to give a summary of the procedure described in the passage. Copy it and complete the blanks.

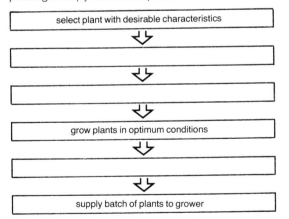

```
┌─────────────────────────────────────────────┐
│   select plant with desirable characteristics │
└─────────────────────────────────────────────┘
                      ⇩
┌─────────────────────────────────────────────┐
│                                              │
└─────────────────────────────────────────────┘
                      ⇩
┌─────────────────────────────────────────────┐
│                                              │
└─────────────────────────────────────────────┘
                      ⇩
┌─────────────────────────────────────────────┐
│        grow plants in optimum conditions      │
└─────────────────────────────────────────────┘
                      ⇩
┌─────────────────────────────────────────────┐
│                                              │
└─────────────────────────────────────────────┘
                      ⇩
┌─────────────────────────────────────────────┐
│        supply batch of plants to grower       │
└─────────────────────────────────────────────┘
```

c) Name TWO environmental conditions given in the passage that would be kept at optimum levels to promote growth.
d) State TWO precautions that are taken to prevent the young plants from becoming diseased.
e) One of the benefits gained by the commercial grower mentioned in the passage is that there is no need to return again and again to pick the fruit crop.

Identify the phrase that tells you why one visit to the crop at harvest time will be enough.
f) What word given in the passage means 'producing a population of identical plants'?
g) What could cause a grower to lose his entire crop? Explain why.

5 Draw up a table to show clearly the differences between the eight fruits shown in the diagram.

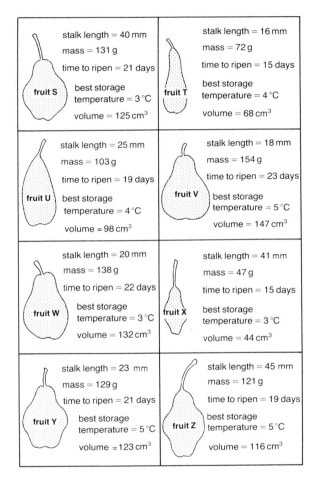

fruit S
stalk length = 40 mm
mass = 131 g
time to ripen = 21 days
best storage temperature = 3 °C
volume = 125 cm^3

fruit T
stalk length = 16 mm
mass = 72 g
time to ripen = 15 days
best storage temperature = 4 °C
volume = 68 cm^3

fruit U
stalk length = 25 mm
mass = 103 g
time to ripen = 19 days
best storage temperature = 4 °C
volume = 98 cm^3

fruit V
stalk length = 18 mm
mass = 154 g
time to ripen = 23 days
best storage temperature = 5 °C
volume = 147 cm^3

fruit W
stalk length = 20 mm
mass = 138 g
time to ripen = 22 days
best storage temperature = 3 °C
volume = 132 cm^3

fruit X
stalk length = 41 mm
mass = 47 g
time to ripen = 15 days
best storage temperature = 3 °C
volume = 44 cm^3

fruit Y
stalk length = 23 mm
mass = 129 g
time to ripen = 21 days
best storage temperature = 5 °C
volume = 123 cm^3

fruit Z
stalk length = 45 mm
mass = 121 g
time to ripen = 19 days
best storage temperature = 5 °C
volume = 116 cm^3

6 Read the passage and answer the questions based on it.

Although often confused with the coconut palm, the cocoa (cacao) tree is an entirely different plant. When mature it is an evergreen tree with large leaves and pinkish-yellow flowers which develop into pods. Each pod contains about forty cocoa beans which are full of useful foodstuffs.

On harvesting, the beans must be fermented, dried, roasted and crushed. Then their contents can be made into highly nourishing cocoa powder (drinking chocolate). Milk chocolate is a mixture of cocoa powder, sugar and milk. The main producers of cocoa are the West Indies and Central and South America.

a) Choose a suitable title for the above passage from the following:

A   Types of useful bean plant
B   Palm oil from coconuts
C   Production of cocoa powder
D   Chocolate for a balanced diet

b) Identify TWO phrases in the passage which tell you that the cocoa plant is of importance in human nutrition.
c) The average annual yield of a cocoa tree is found to be 25 pods. If the total number of beans from 25 healthy pods weigh 1000 g, what is the average mass of one fresh cocoa bean?

Extra Questions

d) If only 0.4 of a cocoa bean's mass is converted into cocoa powder, calculate in kilograms the average yield of cocoa powder for a plantation of 200 trees where a quarter of the trees are too young to produce pods.
e) In a diseased plantation, the total number of beans from 25 pods only weighed 505 g. Express this as a percentage of the weight of beans from 25 healthy pods.

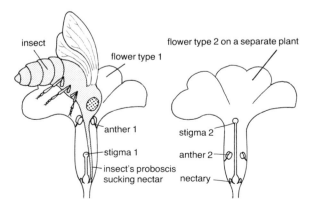

7

Cross-pollination is the transfer of pollen from the anther of one flower to the stigma of another flower on a separate plant of the same species.
a) Explain the difference between cross- and self-pollination.

Extra Question

b) With reference to all of the named structures in the accompanying diagram, describe how cross-pollination takes place between flower types 1 and 2.

8 The following table gives the optimum, maximum and minimum temperatures at which the grains or seeds of three types of plant germinate.

plant	optimum (°C)	minimum (°C)	maximum (°C)
maize	37–44	8	44
wheat	25–30	3	35
cucumber	31–37	15	49

a) How many of the plants could be germinated at 12 °C?
b) Which plants could be germinated at 47 °C?

Extra Question

c) Could all three plant types be germinated within their optimum range of temperature simultaneously in the one greenhouse? Explain your answer.

9 The diagram opposite shows six methods of asexual reproduction in plants.
a) From the information in the diagram state TWO differences between a bulb and a corm.
b) Identify ONE feature shared by both a tuber and a rhizome.

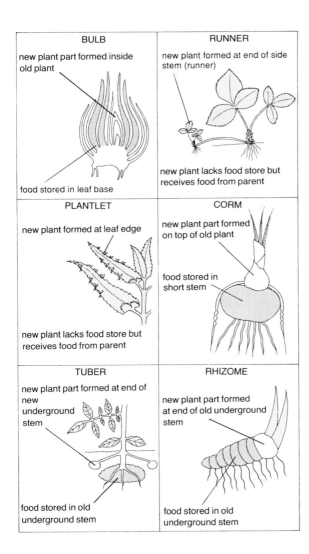

BULB
new plant part formed inside old plant

food stored in leaf base

RUNNER
new plant formed at end of side stem (runner)

new plant lacks food store but receives food from parent

PLANTLET
new plant formed at leaf edge

new plant lacks food store but receives food from parent

CORM
new plant part formed on top of old plant

food stored in short stem

TUBER
new plant part formed at end of new underground stem

food stored in old underground stem

RHIZOME
new plant part formed at end of old underground stem

food stored in old underground stem

Extra Question ⊞

c) Construct a key of paired statements that allows each method of asexual reproduction to be correctly classified. (Use only the information given.)

10 One method of natural vegetative propagation amongst plants is bulb formation. As part of an investigation of bulbs, a girl was given an onion bulb and asked to test its fleshy storage leaves for simple sugar and starch.

She found sugar to be present and starch to be absent.

She concluded that all bulbs store sugar but not starch in their leaf bases.

a) Why was she not justified in drawing this conclusion?

Extra Question ⊞

b) What information would the girl need to have before attempting to draw a conclusion about the type of food stored in bulbs?

11 Making food

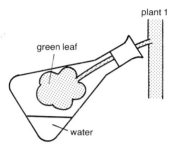

plant 1

green leaf

water

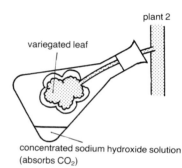

plant 2

variegated leaf

concentrated sodium hydroxide solution
(absorbs CO_2)

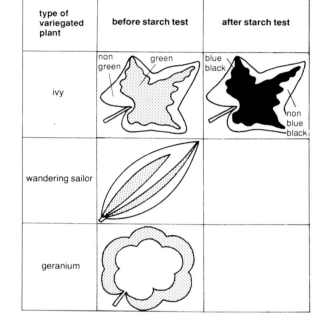

type of variegated plant	before starch test	after starch test
ivy	non green · green	blue black · non blue black
wandering sailor		
geranium		

1 The above experiment is an alternative method of demonstrating that carbon dioxide is needed by a green plant for photosynthesis.

State TWO ways in which this experimental set-up would have to be altered to make it fair.

2 A variegated ivy plant was given all of the conditions needed for photosynthesis. When one of its leaves was tested for starch, the result shown in the table opposite was obtained. It was therefore concluded that chlorophyll is required for photosynthesis.

a) Copy and complete the table to show the appearance of the other two leaf types under the same conditions after the starch test.

b) Imagine that a cutting from the wandering sailor plant had been kept in darkness for two days. Draw a labelled diagram of how one of its leaves would appear after the starch test.

3 A scientist grew specimens of a species of cereal plant at different light intensities for several hours and then measured the amount of sugar produced at each light intensity as shown in the following table.

light intensity (foot candles)	sugar production (g/kg of dry leaves)
0	0
100	1.7
200	3.3
300	4.8
400	5.7
500	6.0
600	6.0
700	6.0

a) Using graph paper, construct a line graph of the above results. (Use the horizontal axis for light intensity and the vertical axis for sugar production.)

b) Express the amount of sugar produced per kilogram of dry leaves at light intensity 700 foot candles as a percentage of the dry mass of leaves.

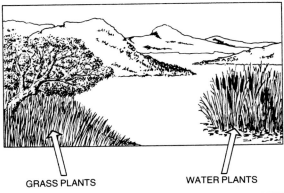

GRASS PLANTS　　　　　　　WATER PLANTS

time on 24 hour clock	CO₂ conc. in air between leaves
10.00	high
13.00	low
16.00	medium

time on 24 hour clock	CO₂ conc. in water between leaves
10.00	high
13.00	medium
16.00	low

4 It was concluded from the results in the above experiment that grass plants were absorbing more CO_2 at 13.00 hours than at either of the other two times.
a) At which of the three times was photosynthesis occurring at the greatest rate in water plants?
b) The pH value of a liquid is known to decrease as its CO_2 concentration increases. At which of the three times will the pond water be at the lowest pH?

5 In an experiment to investigate water loss in two types of plants, leaves were attached to simple balances as shown in the following diagrams.

a) Which leaf lost more water in experiment 1?
b) Which leaf lost more water in experiment 2?
c) What can be concluded about the location of most or all of the stomata in a waterlily leaf from the above experiment?

6 The following graph shows the effect of increasing light intensity on the uptake of carbon dioxide by a green plant.

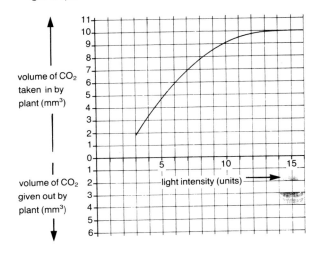

a) How many units of light was the plant receiving when it took in 6 mm³ of CO_2?
b) What volume of CO_2 was taken in by the plant at a light intensity of 7 units?
c) What volume of CO_2 was taken in by the plant at a light intensity of 14 units?

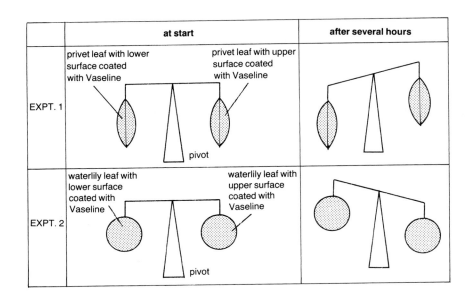

	at start	after several hours
EXPT. 1	privet leaf with lower surface coated with Vaseline　　privet leaf with upper surface coated with Vaseline　　pivot	
EXPT. 2	waterlily leaf with lower surface coated with Vaseline　　waterlily leaf with upper surface coated with Vaseline　　pivot	

Extra Question ⊞

d) With the aid of a ruler work out the light intensity at which CO_2 is neither taken in nor given out by the plant.

---

7 The table below gives the results from an investigation into the possible relationship between monthly rainfall and photosynthesis (measured as dry mass of a standard number of leaf discs).

month	rainfall (cm)	dry mass of leaf discs (g)
January	20	2.5
February	14	3.0
March	12	3.5
April	10	4.0
May	8	4.0
June	5	4.5
July	3	5.0
August	4	5.0
September	8	4.5
October	14	4.0
November	19	3.0
December	20	2.0

a) Name the three driest months of the year
b) How many months were wetter than April?
c) During which months did most photosynthesis occur?

Extra Question ⊞

d) Present the data graphically by first drawing a bar graph of rainfall and then a line graph of dry mass of leaf discs using a common horizontal axis to highlight the possible relationship.

Extra Question ⊞

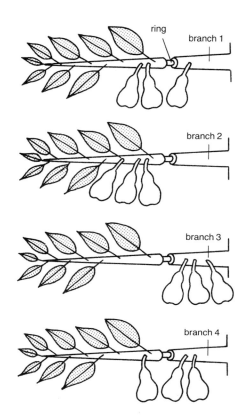

8 The diagrams show four branches of a pear tree at the start of a ringing experiment where the fruit grower is attempting to produce unusually large pears.
   The table below shows the total mass of the three pears per branch several months later. Copy the table and complete it by matching the four sets of results with the four branches.

total mass of three pears per branch	branch number
75 g	
300 g	
525 g	
750 g	

12 The need for food

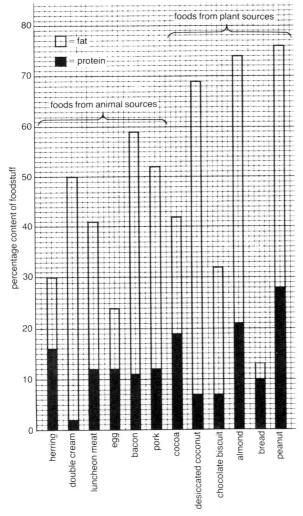

c) Name the animal foods that contain a lower percentage of fat than pork.
d) How many foods contain more fat than protein?
e) Which of the following contains least protein? herring, cocoa, almond, peanut
f) Name THREE foods that contain an equal percentage of fat.

2 a) Calculate the average fat content for the foods shown in the table below.

food	fat content (g/kg)
butter	810
cheese	345
herring	185
lamb	302
milk	38

b) Calculate the average protein content for the foods shown in the table below.

food	protein content (g/kg)
beans (baked)	5.1
beef	18.2
cheese	25.5
milk	3.3
peanuts (roasted)	28.2
peas	5.8
potatoes	2.1

1 Using only the information given in the above bar graph, answer the following questions.
a) Which THREE animal foods contain the same percentage of protein?
b) Which plant foods contain a higher percentage of protein than bread?

3 A human's first set of teeth are called milk or deciduous teeth. The first milk teeth to appear through the gums are the incisors at about 6–10 months after birth. The full set is normally present by about 2 years but always lacks molar teeth. The first permanent teeth (of the second set) appear at the age of about 6 years. These are the front molars. The milk teeth are then replaced by further permanent teeth. Incisors grow in followed by the premolars and then the canines. Later, at about 12 years, the second molars appear and finally the third molars ('wisdom' teeth) at around 18 years or even later giving a full set of 32 teeth.

a) What is the main theme of the above passage?
b) Identify TWO differences between first and second sets of teeth.
c) List the order in which the permanent teeth appear.
d) How many teeth are present in a full set of milk teeth?

4 The following diagram charts the progress of various foodstuffs along the human gut (alimentary canal).

type of food at start	in mouth	in stomach	in small intestine	substance(s) present after digestion
fat				fatty acids + glycerol
starch				sugar
vitamins				vitamins
protein				amino acids

KEY  = undigested food   = enzyme-digested food

a) Which type of food remains undigested as it passes through the alimentary canal?
b) Which type of food undergoes digestion in both mouth and small intestine?
c) Name the type of food that is digested in the small intestine only.
d) How many different types of food undergo digestion in the stomach?

5 A small sample of the natural liquid present in a pig's gut was taken from each of the four lettered regions indicated in the diagram.

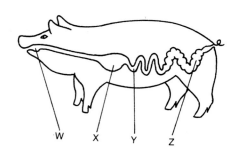

Each sample was put into a small hole in a dish of starch agar. After 24 hours at 37 °C, the dish was flooded with iodine solution. The result is shown in the diagram.

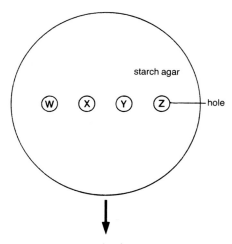

starch agar

W   X   Y   Z ——— hole

result after flooding with iodine solution

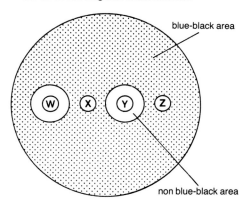

blue-black area

W   X   Y   Z

non blue-black area

a) What colour results when iodine solution reacts with starch?
b) What conclusion can be drawn from the experiment about a pig's ability to digest starch in regions W, X, Y and Z of its gut?
c) Explain why starch digestion is possible in some of these regions of the gut but not in others.

Extra Questions

6 Choose suitable forms of calculation and then copy
   and complete the final column in the following table.

vitamin	minimum daily requirement (mg)	notes	least amount of food (g) that will supply minimum daily requirement of vitamin
A	0.75	166 g margarine contains 1.5 mg of vitamin A	
B_1	1.2	525 g wheat germ contains 3.6 mg of vitamin B_1	
B_{12}	0.003	800 g cheese contains 0.012 mg of vitamin B_{12}	
C	30	100 g lettuce contains 15 mg of vitamin C	
D	0.0025	50 g sardines contains 0.0025 mg of vitamin D	

7 The following table gives a summary of a series of
   experiments in which tough meat was treated with
   the juice squeezed out of the pawpaw fruit (which
   grows on palm-like trees in tropical countries).

	age of fruit		state of fruit		pH conditions		resultant state of meat
	mature	immature	boiled	unboiled	neutral	acidic	
experiment 1	✓		✓		✓		tough
2		✓	✓		✓		tough
3	✓			✓	✓		tough
4		✓		✓	✓		tender
5	✓		✓			✓	tough
6		✓	✓			✓	tough
7	✓			✓		✓	tough
8		✓		✓		✓	tough

Study the information carefully then construct a
hypothesis to account for the results.

13 Reproduction

1 Imagine that you are the parent of a three-month-old baby girl who receives a feed every 2–3 hours. The baby has a runny nose and has been coughing but not crying much. Her temperature is 37 °C and her faeces are normal but she has just brought up her entire feed for the first time. When you look up the family medical book for help you find the following chart.

a) What seems to be wrong with the baby?
b) Imagine that you have decided to take the advice given in the flow chart. Describe what you would do over the next few hours.
c) If you read a flow chart like this one but fail to understand it or fail to find a satisfactory answer, what should you do?

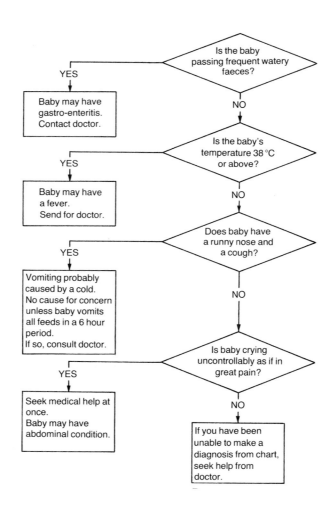

2 Read the passage and answer the questions that are based on it.

A few hours after birth, a human baby starts to suckle at the mother's breast or from a bottle. The mother's milk contains everything that the baby needs except iron. A store of this chemical element has already been obtained from the mother's bloodstream by the foetus during gestation. This supply lasts for a few months after birth.

Bottle-fed babies receive dried cow's milk which has been altered in advance by the manufacturers since natural cow's milk contains more protein and less vitamin A and sugar than human milk. If strict hygiene is not observed, bottle feeding can pass germs on to the baby. Human milk is free of germs and in addition contains antibodies which defend the baby against diseases. Dried cow's milk lacks these human antibodies.

The crying response in babies is present from birth. It consists of muscular activity, exaggerated breathing, watering of the eyes and high pitched sounds. Almost all young mammals cry when insecure, frightened or in pain. By crying the young animal signals to its parents that something may be wrong. If it continues to cry, the parent will normally examine the young animal for sources of pain or injury.

a) Give TWO changes that manufacturers have to make to cow's milk to make it suitable for human consumption.

b) From where does a human baby obtain iron during the first few months after birth?

c) In addition to containing exactly the right balance of foodstuffs, state TWO other ways in which human milk is better than cow's milk for infants.

d) State the THREE situations given in the passage that can make a baby cry.

e) Name the main signal present in the crying response that quickly draws the parents' attention to the infant.

3 The following diagram gives six pieces of information about each of nine different animals native to North Island, New Zealand. Draw up a table to display this information clearly.

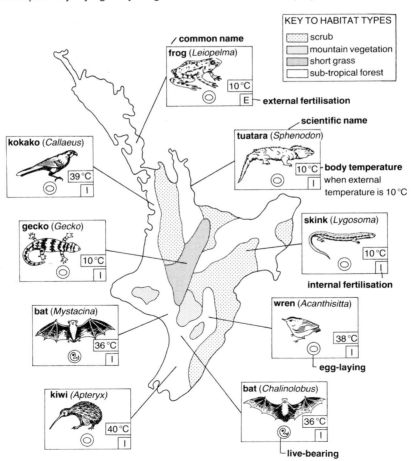

KEY TO HABITAT TYPES
- scrub
- mountain vegetation
- short grass
- sub-tropical forest

common name

frog (*Leiopelma*) 10 °C E — **external fertilisation**

scientific name

tuatara (*Sphenodon*) 10 °C — **body temperature** when external temperature is 10 °C

kokako (*Callaeus*) 39 °C I

skink (*Lygosoma*) 10 °C — **internal fertilisation**

gecko (*Gecko*) 10 °C I

bat (*Mystacina*) 36 °C I

wren (*Acanthisitta*) 38 °C I — **egg-laying**

kiwi (*Apteryx*) 40 °C I

bat (*Chalinolobus*) 36 °C I — **live-bearing**

4 The following graph represents the results from a survey done to investigate the number of twin births that occur per 1000 pregnancies for women of different ages living in Britain.

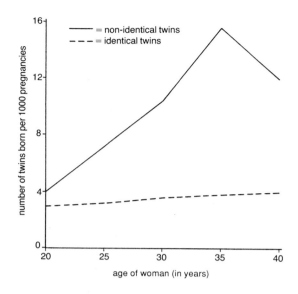

a) Suggest why as many as 1000 pregnancies were studied in the survey.
b) Women from all walks of life (i.e. socio-economic groups) were included. Why?
c) By comparing the line graphs, draw TWO appropriate conclusions from the results.

Extra Questions

5 100 eggs are present in each clutch laid by a green turtle. Assume that she lays 4 clutches at 9-day intervals on the same beach. Monitor lizards then dig up and eat 40% of the eggs. All the remaining eggs hatch out but 80% of the young turtles are caught by gulls and fiddler crabs as they make their way across the beach. Calculate how many baby turtles reach the sea.

6 A shepherd recorded the average mass of the lambs in his flock in the spring of 1984. Some ewes only gave birth to a single lamb which they reared themselves. Some gave birth to twins. Of these ewes, some were left to rear both of their twins normally; others were allowed to rear a 'single' twin and a ewe with no lambs was given the other 'single' twin to rear.

Here are some extracts from the shepherd's diary.

1st March
Many lambs born to-day.
Average mass of single lamb 6 kg.
Average mass of twin lamb 4·5 kg.

29th March
Lambs all doing well despite snow.
Average mass of single lamb 15 kg,
normal twin 10·5 kg and 'single' twin 12·5 kg.

26th April
Snow away and all lambs surviving.
Average mass of single lamb 24 kg,
normal twin 17·5 kg and 'single' twin 21 kg.

24th May
Weather much warmer. Clear differences in size amongst lambs.
Single lamb now 33 kg, normal twin 24·5 kg and 'single' twin 29 kg on average.

a) Present this information in a table so that the average mass for each type of lamb as it changes with time can be followed at a glance.
b) Draw TWO conclusions from the information in your table.

14 Water and waste

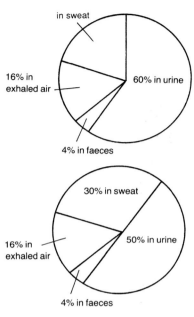

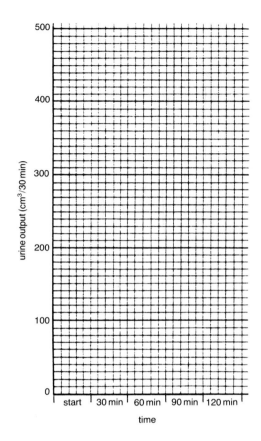

1 The first pie chart shown represents a person's daily water loss.
a) What percentage of this person's water was lost as sweat?
b) If the total volume of water lost was 2500 cm³, what volume of water was lost in exhaled air?
The second pie chart represents the same person's daily water loss at a different time of year.
c) Suggest what environmental factor was responsible for the difference between the two pie charts. Explain your answer.

2 The results given in the following table are from an experiment to investigate the response of a man to drinking one litre of water. Urine was collected at half-hourly intervals.

time (min)	urine output (cm³/30 min)
start (drink given)	50
30	340
60	490
90	160
120	40

Using the axes given opposite and similar graph paper, present the above data as a bar graph.

3 The following tables show the water balance figures for a human subject who remained under constant environmental conditions for two days.

water gain	day 1	day 2
in food	850 cm³	850 cm³
in drink	1300 cm³	1600 cm³
formed during respiration	350 cm³	350 cm³
total gain	2500 cm³	2800 cm³

water loss	day 1	day 2
from lungs		
from kidneys	1500 cm³	
from skin	500 cm³	500 cm³
in faeces	100 cm³	100 cm³
total loss	2500 cm³	2800 cm³

a) In what way did water gain differ between day 1 and day 2?
b) Calculate how much water was lost in exhaled air during day 1.
c) Calculate the volume of urine expelled during day 2.

Extra Questions

⊞

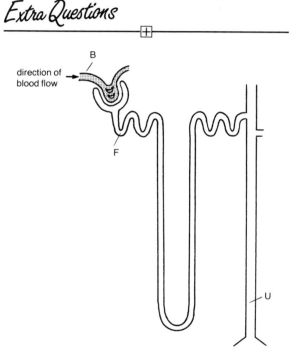

direction of blood flow

B

F

U

4 The accompanying diagram shows a kidney filtering unit (nephron) and part of its blood supply. The concentrations of some of the substances found in the blood (B), the filtrate (F) and the urine (U) are given in the following table.

	urea (g/100 cm³)	glucose (g/100 cm³)	protein (g/100 cm³)	salts (g/100 cm³)
blood (B)	0.03	0.10	8.00	0.72
filtrate (F)	0.03	0.10	0.00	0.72
urine (U)	2.00	0.00	0.00	1.50

a) Which substance did not pass from B to F?
b) Which substance present in the filtrate was completely reabsorbed back into the bloodstream?
c) Which substance became concentrated in urine to the greatest extent?
d) If a man produced 180 litres of filtrate and 1.8 litres of urine in one day, what percentage of filtrate was passed as urine?

5 Read the passage and answer the questions which are based on it.

In addition to *diabetes mellitus* (sugar diabetes), there exists another disease called *diabetes insipidus*. This is characterised by excessive thirst and the production of large quantities of very dilute urine. The disease can be caused by a malfunction of the posterior lobe of the pituitary gland. This results in insufficient ADH (anti-diuretic hormone) being secreted into the bloodstream and therefore less water being reabsorbed from the kidney tubules back into the blood. Instead vast quantities of dilute urine are produced.

In some cases *diabetes insipidus* is treated by taking ADH in the form of nasal snuff. Without treatment a sufferer has to drink many litres of water in a single day to make up for the water losses.

a) Explain why a malfunction of the posterior lobe of the pituitary gland can cause *diabetes insipidus*.
b) Apart from thirst, what is the main symptom of *diabetes insipidus*?
c) Why does lack of sufficient ADH lead to the symptom you gave as your answer to (b)?
d) Suggest what could happen if a sufferer of *diabetes insipidus* took an exceptionally high dose of nasal snuff containing ADH.

15 Responding to the environment

1 20 woodlice were released into a choice chamber. The numbers of animals present in both sides were recorded every minute for 10 minutes. The results are shown in the following graph.

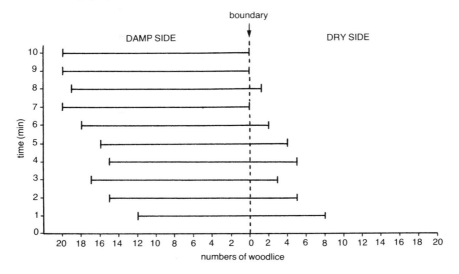

a) What was the one variable factor investigated in this experiment?
b) How many woodlice were present on the (i) dry side at minute 5? (ii) damp side at minute 8?
c) At which minute were there (i) 17 woodlice on the damp side? (ii) 2 woodlice on the dry side?
d) What conclusion can be drawn from this experiment about the response of woodlice to the environmental factor under investigation?

e) Why must all the woodlice used be of the same species?
f) In a further experiment the boundary between the damp and dry sides of the choice chamber was blocked with cardboard after 1 minute. The seven woodlice trapped in the dry side eventually slowed down after several minutes and bunched up together against the side of the chamber in one group. Suggest why this response is of survival value to the animals.

2 The following graphs record the activity of three processes that occur in a human body over a period of three days.
a) For how many hours did the person sleep each night?
b) With reference to cycle of activity, what generalisation can be made about the three body processes represented in the graphs?

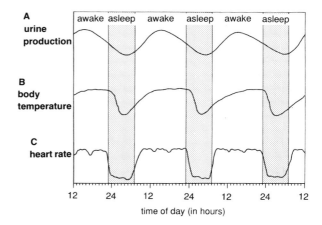

Extra Questions ⊞

c) Suggest why the rhythmical behaviour pattern shown in graph C is of benefit to the person.
d) Cell division in the skin shows the reverse rhythm to that of heart rate. Between which hours (approximately) would you expect wound healing to be most rapid?

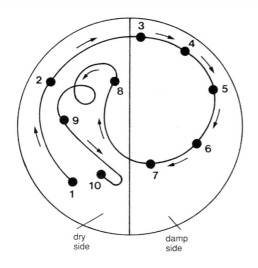

dry side   damp side

3 The above diagram shows a choice chamber into which a boy dropped a woodlouse through a hole in the lid. The woodlouse landed at position 1. (All holes in the choice chamber's lid were kept closed during the experiment.)

The distance between two numbered dots represents the distance covered by the woodlouse every 5 seconds.
a) What was the one variable factor controlled by the experimenter in this investigation?
b) How many conditions of this factor were included?
c) Describe the effect of each of these conditions on the woodlouse's rate of activity.

Extra Questions ⊞

d) Explain how you arrived at your answer to part (c).
e) How long did it take the woodlouse to move from position 1 to position 10?
f) From the results, the boy concluded that all woodlice behave in this way. Why was he NOT justified in making such a generalisation?

4 The following experiment was set up to investigate the effect of two environmental factors on the distribution of a tiny animal that feeds on green algae at the surface of the pond where it lives.

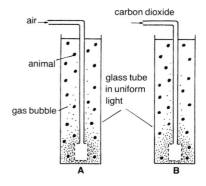

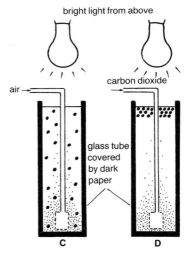

a) Name the one environmental factor by which tubes A and B differ.
b) What effect does this difference have on the distribution of the animals?

Extra Questions

c) Name the one environmental factor by which tubes A and C differ.

d) What effect does this difference have on the distribution of the animals?

e) Consider the distribution of the animals in tube D. What conclusion does this result allow us to draw from the experiment?

f) Suggest why the response shown by this animal to its environment is of survival value.

5 The apparatus below was designed by a group of pupils in order to investigate the response of slugs to a choice between room temperature and near freezing conditions. It was their intention to insert 20 slugs of the same species and then use the apparatus exactly as shown over a period of one hour without making any alterations.

a) Suggest why they made tiny holes in the lemonade bottles.

b) Consider carefully the aim of the experiment and then state a source of error that will affect the intended attempt described above.

c) Suggest how this source of error could be minimised.

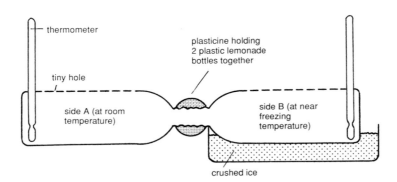

16 Movement

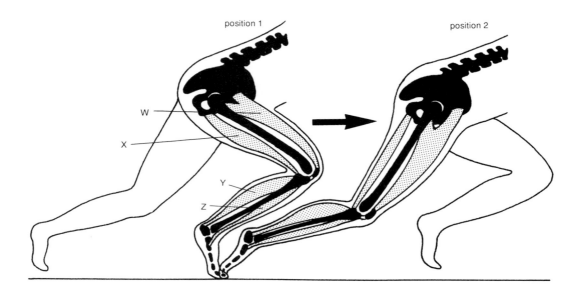

position 1   position 2

1 The above diagram shows two positions of an athlete's right leg during a race. During the change from position 1 to position 2, which muscles in the diagram (i) contract (ii) become relaxed?

2 The human leg can be made to move in many different ways. Six of these movements are shown in the diagram. Each movement, described in the following table, is brought about by a different group of muscles.

a) Match movements 1–6 with muscle groups A–F which bring them about.
b) Each group of muscles in the table is antagonistic to another group also in the table. Identify the three pairs.

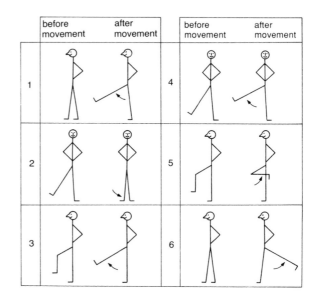

muscle group	description of movement brought about
A	two parts of limb pulled towards each other by closing a hinge joint
B	limb pulled forwards and out by movement at a ball and socket joint
C	limb pulled sideways and outwards away from body
D	limb pulled backwards and out by movement at ball and socket joint
E	two parts of limb pulled away from each other by opening a hinge joint
F	limb pulled sideways and inwards towards body

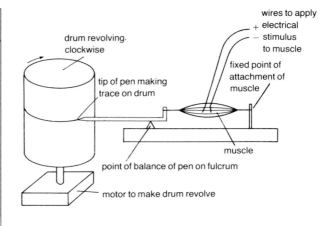

Extra Questions

3 Which of the following graphs best represents the maximum load which can be supported at different arm extensions?

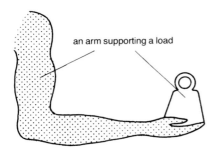

an arm supporting a load

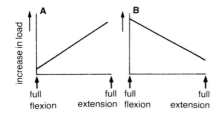

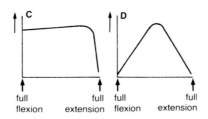

4 When a muscle is removed from an animal (e.g. a frog) and fitted to the apparatus shown above, the muscle can be made to contract by applying an electrical stimulus to it. A record of muscle contraction is produced by the pen making a trace on the revolving drum.

a) In which direction will the tip of the pen move when the muscle (i) contracts, (ii) relaxes?

b) The two traces shown below were obtained by repeating the experiment at two different temperatures.

(i) Draw TWO conclusions from the results.

(ii) State TWO factors that must be kept constant during this experiment to make it fair.

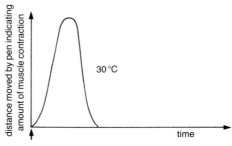

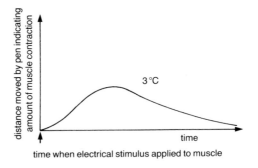

17 The need for energy

1 The table below shows the amount of energy expended per hour by a 15 year-old girl when performing various activities.
    Present the information as a bar chart.

activity	energy used (kJ/hour)
walking	1200
classwork	800
sport	2000
sitting at rest	330
sleeping	270

2 The volumes of air entering and leaving the lungs can be measured using a piece of apparatus in which a pen moves up and down on a revolving drum making a trace as the person breathes in and out. The following graph represents such a trace.

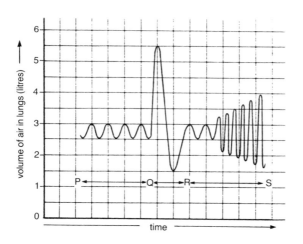

a) In which direction does the pen move when air is being (i) inhaled (ii) exhaled?
b) What volume of air is inhaled and exhaled at each breath during period P–Q when the subject is at rest?

Extra Questions

c) Describe in detail the breathing pattern of the subject during period Q–R on the graph.
d) State the total lung capacity of this person as indicated by the graph.
e) State ONE factor which could account for the trace recorded during period R–S on the graph.

3 The following bar graph shows the rate of blood flow to various parts of a student's body under differing conditions of exercise.

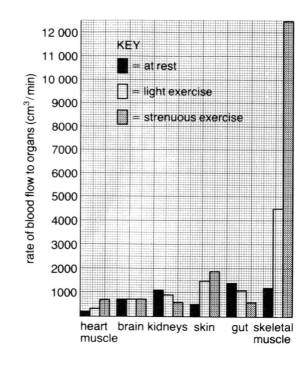

a) Using this information, copy and complete the following table.

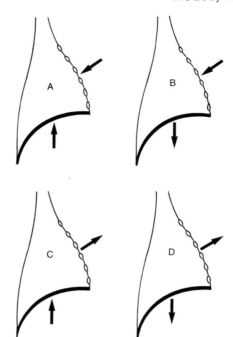

part of body	rate of blood flow (cm³/minute)		
	at rest		strenuous exercise
heart muscle		350	750
	750	750	
kidneys		900	600
	500		1900
gut	1400	1100	
skeletal muscle	1200		

b) What effect does increasingly strenuous exercise have on rate of blood flow to (i) brain, (ii) skeletal muscle, (iii) gut?
c) Which other part(s) of the body shows the same trend in response to increase in exercise as (i) gut, (ii) heart muscle?

Extra Questions

d) Suggest a way in which the appearance of the skin would change as a result of strenuous exercise. Explain why.
e) Calculate the **total** volume of blood per minute being pumped by the left ventricle to all of the parts of the body given in the graph during a period of (i) rest, (ii) light exercise, (iii) strenuous exercise.
f) If the student's pulse rate at rest is 65 beats/minute, what volume of blood is being pumped out of the left ventricle during each heart beat?

4 The accompanying diagrams show four simplified outlines of the human chest cavity. In which diagram do the arrows correctly indicate the movements of the diaphragm and rib cage during (i) inspiration, (ii) expiration?

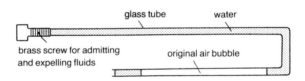
glass tube   water
brass screw for admitting and expelling fluids   original air bubble

5 The above apparatus was used to investigate the composition of a bubble of exhaled air after exercise. With the bubble always at constant temperature and pressure, the following results were obtained:

length of gas bubble in contact with water = 100 mm
length of bubble after contact with potassium hydroxide (which removes CO_2) = 95 mm
length of bubble after contact with potassium pyrogallol (which removes oxygen) = 80 mm

a) Calculate the percentage of CO_2 present in the original bubble
b) Calculate the percentage of oxygen present in the original bubble.

6 In the diagram below, a capillary is in close contact with structure R. The relative concentrations of carbon dioxide (CO_2) and oxygen (O_2) are given at three different sites.

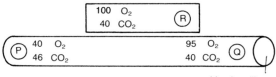

blood capillary

Which one of the following is correct?

A R is an alveolus and blood flow is from P to Q.
B R is an alveolus and blood flow is from Q to P.
C R is a muscle and blood flow is from P to Q.
D R is a muscle and blood flow is from Q to P.

7 The graph opposite shows the pressure changes that occur in the left ventricle and the aorta during one heart beat.
a) What is the highest pressure recorded for the left ventricle?
b) Contraction of the left ventricle is represented by the part of the graph lettered: A–B   B–C   D–E (Choose correct answer.)
c) (i) At which lettered point on the graph does ventricular pressure first become equal to the pressure in the aorta?   P   Q   R   (Choose correct answer.)

(ii) After this point on the graph, for a short time ventricular pressure slightly exceeds aortic pressure. Which valve(s) will be open during this period?
(iii) At which lettered point on the graph will these valves start to close?   P   Q   R   (Choose correct answer.)

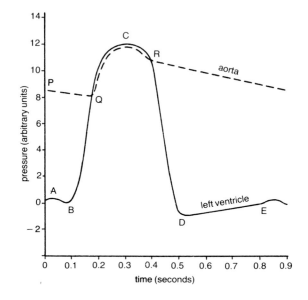

18 Co-ordination

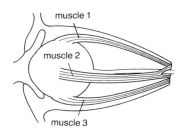

muscle 1
muscle 2
muscle 3

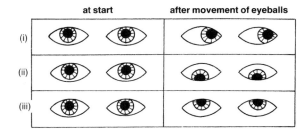

	at start	after movement of eyeballs
(i)		
(ii)		
(iii)		

1 The upper diagram gives a view of the human eyeball from the side. Movements of the eyeball are controlled by six muscles, of which three are shown in the diagram.
a) Which muscle (attached to each eyeball) must contract to make the eyes perform movements (i), (ii), (iii) shown in the above table?

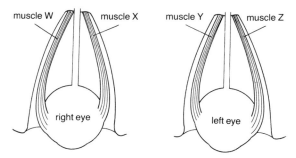

muscle W   muscle X   muscle Y   muscle Z

right eye   left eye

The above diagram gives a view of both eyeballs from above. Only two of the muscles controlling eyeball movements are shown for each eyeball.
b) With reference to these muscles only, explain how the movement of the eyeballs shown below is brought about.

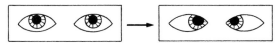

2 When a boy, sitting in a darkened room, had covered his left eye, a bright torch was switched on for 20 seconds and then the diameter of the pupil of his right eye was measured. This was called trial 1. The torch was switched off for 20 seconds and during this time moved to a position further away from the eye. It was then switched on again for 20 seconds and the diameter of the pupil measured (i.e. trial 2). This procedure was repeated for two more trials and then four trials were done bringing the torch closer and closer to the eye each time.
   The results are presented in the following bar chart.

a) Between which two trials did the following changes in diameter of the pupil occur?
(i) biggest increase, (ii) smallest increase,
(iii) biggest decrease, (iv) smallest decrease.
b) At which trial was the torch (i) furthest away from the eye? (ii) nearest to the eye?
c) At what two trials was the torch at the same distance from the eye?
d) In general, what relationship exists between the diameter of the pupil and the distance of the torch from the eye?

Extra Questions

3 Each cerebral hemisphere of the human brain has a region called the sensory area and it is here that sensations such as touch and pain are perceived. Each part of the body capable of sending such impulses to the brain is represented by an area on the sensory region. However, the area of the brain

devoted to each body part is found to be in proportion not to the actual size of the body part but instead to the relative number of sensory receptors present in that body part.

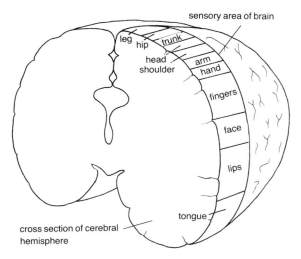

cross section of cerebral hemisphere

The above diagram indicates how much of the brain's sensory area is given to each body part. The diagram below shows an imaginary human figure ('sensory homunculus') whose body parts have been drawn in relation to their sensitivity as opposed to their actual size.

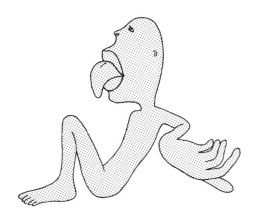

a) Name TWO types of sensation perceived by the sensory area of the cerebrum.
b) Account for the fact that a leg is a big part of a normal human body yet it is represented by a fairly small area on the cerebrum's sensory region.
c) Which body part contains most sense receptors relative to its actual size?
A trunk   B shoulder   C hip   D tongue

d) Which of the following structures has fewest nerve endings in relation to its actual size?
A arm   B face   C fingers   D lips
Each cerebral hemisphere also possesses a motor region which controls body movements. Here the size of the brain part allocated to each body part is related not to the body part's size but to its degree of mobility as illustrated by 'motor homunculus' shown below.

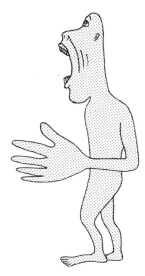

e) Identify TWO especially mobile parts of the body.
f) Which third part of the body is sufficiently mobile to be used, with much practice, to operate a pencil or paint brush?
g) With reference only to the pinna (ear flap), predict how 'motor homunculus' would differ if a rabbit had been drawn. Explain your answer.

4 A reflex action is a simple act of behaviour whose function is protective. In each case the reflex action is a rapid, automatic response to a stimulus. For example, when an object touches the eye the eyelid muscle contracts, bringing about blinking which prevents damage to the eye. When foreign particles such as pepper enter the nasal tract, a sudden contraction of the chest muscles makes the person sneeze and remove the unwanted particles from the nose. Although reflex actions are involuntary, some can be partly altered by voluntary means; the person can to a certain extent resist blinking or sneezing.

Some reflex actions cannot, however, be altered by voluntary means. When food is present in the gut, muscles in the gut wall contract bringing about peristalsis. This ensures efficient digestion by mixing the food thoroughly with digestive enzymes. In dim light the pupil of the eye becomes dilated (enlarged)

reflex action	stimulus	response	protective function	can be partly altered by voluntary means?
blinking		contraction of eyelid muscle		yes
	presence of food in gut		ensures movement and therefore efficient digestion of food	no
	foreign particles in nasal tract	sudden contraction of chest muscles		
dilation of eye pupil			improves vision in poor lighting	

as a result of movement of the iris muscle. This reflex action improves the person's vision in poor lighting. Peristalsis and pupil dilation cannot be partly resisted or prevented by voluntary means.

Copy and complete the above table using the information given in the passage.

5 The lens of the human eye is surrounded by a ring of muscle fibres called ciliary muscle. This muscular ring is attached to the edge of the lens by threadlike suspensory ligaments as shown in the following diagrams.

In the upper eye the ring of ciliary muscle has become relaxed, increasing the diameter of the circle that it forms. This makes the suspensory ligaments become taut and pull the lens out into a wide, thin shape which is ideal for bringing to a focus rays of light from a distant object.

To focus the rays of light coming from a near object, the shape of the lens has to be changed as shown in the lower diagram.

a) What shape has the lens become in order to focus light from a near object?
b) Describe in your own words how this change has been brought about.

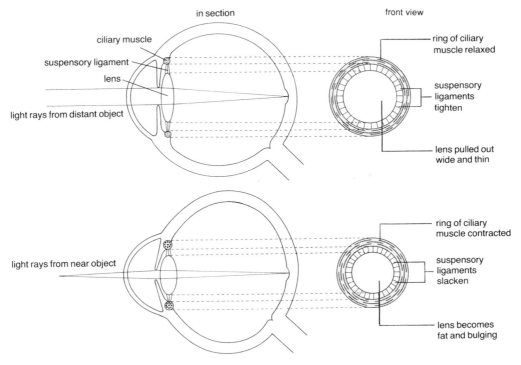

6 When a person moves from bright light into almost total darkness, he is temporarily blinded, but after a few minutes the rod cells in each retina respond and he can see fairly well. His eyes are said to be dark-adapted. If he now returns to bright light he is briefly dazzled until the cone cells in each retina respond. Then he can see properly again and his eyes are said to be light-adapted.

The degree of sharpness of detail seen by an eye is called its visual acuity. This is compared for a dark-adapted and light-adapted eye in the following graph.

a) Explain the difference between a light-adapted and a dark-adapted eye.

b) (i) in which type of eye is the relative acuity of vision greatest at (1) the fovea, (2) the side of the eye?

(ii) Relate this difference to the distribution of rods and cones in the retina.

c) Acuity of vision for both types of eye is zero at region X on the retina. Suggest why.

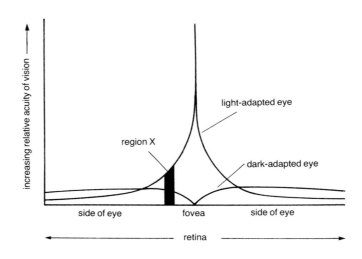

19 Changing levels of performance

1 The table gives information about the average energy output of a British soldier when performing various activities before and after three months of training.
a) Construct a bar chart to illustrate all the information in the table.
b) Which activity was unaffected by training?
c) In general what effect did training have on all the other activities?

	energy output (kJ/min)	
activity	before training	after training
guard duty	10.9	10.7
ironing uniform for inspection	17.2	16.5
cleaning kit and rifle	11.3	11.3
assault course	41.2	35.7
slow marching	15.5	14.0
quick marching	22.7	20.3

2 The following graph refers to the pulse rates of two students X and Y. X trains regularly but Y does not undertake regular physical exercise.

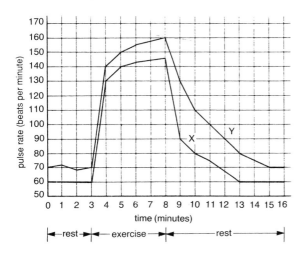

a) State the resting pulse rate for each student.
b) What was the pulse rate for each student one minute after exercise came to a halt?

c) From the graph, state TWO effects of training on pulse rate.

3 The steps in the procedure used to measure length of recovery time after vigorous exercise are given below but in the wrong order.
(i) Exercise vigorously for three minutes (e.g. doing step test).
(ii) Calculate time required for pulse rate to return to normal.
(iii) Measure normal pulse rate before starting to exercise.
(iv) Take pulse rate at one minute intervals until normal pulse is recorded.
a) Arrange steps (i)–(iv) in the correct order.
b) One afternoon this procedure was used to compare the recovery times of several college students. Some students had been playing sports at lunchtime and had missed lunch. Some students had been studying until late the night before while others had gone to bed early.
(i) Explain why the investigation is not a fair test.
(ii) Suggest how it could be adapted to make it fair.

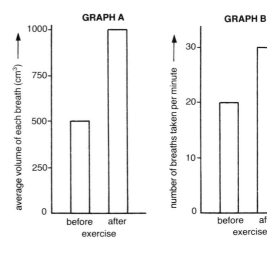

4 The above bar graphs show the effect of exercise on a girl's rate and depth of breathing.
a) Which graph shows (i) rate, (ii) depth of breathing?
b) What effect does exercise have on both rate and depth of breathing?

Extra Questions

c) (i) What total volume of air passes in (and out) of the girl's lungs each minute after exercise?
(ii) By how much is this volume greater than that passing in (and out) of her lungs before exercise?
d) The air that the girl was breathing was found to contain 150 particles (pollen grains and fungal spores) per 500 cm^3 of air.
(i) Calculate how many particles she would inhale per hour when not exercising.
(ii) Explain why very few of these particles would reach her alveoli.

5 The information in the following table refers to an alert, sober driver in a well-maintained car with good brakes and tyres.

a) Copy and complete the table.
b) In general, predict the effect of drinking four pints of beer on a driver's reaction time.
c) Describe how this would affect his overall stopping distance at any of the speeds given in the table.

6 The Harvard step test is often used to measure fitness. The subject steps onto and down from a 500 mm high platform 30 times per 5 minutes or until fatigue forces him to stop. His pulse rate is taken for the period exactly 1 to 1½ minutes after exercise and his fitness index (FI) calculated using the formula:

$$FI = \frac{\text{total exercise time (in seconds)} \times 100}{5.5 \times \text{pulse rate}}$$

The following table shows the categories of fitness.

FI	fitness category
less than 50	poor
50–80	average
more than 80	good

The pulse rates of John, Joe and Jim were found to be 90, 120 and 65 respectively during the period 1 to 1½ minutes after each had completed the 5 minute step test. To which fitness category does each boy belong?

car speed (miles/hour)	B R A K E S A P P L I E D	shortest stopping distance (in feet)		
		thinking distance	braking distance	overall stopping distance
20		20	20	40
30		30	45	75
40		40		120
50			125	175
60		60		
		70	245	315

Section 6 Inheritance

20 Variation

1 After a visit to a zoo, 2000 young schoolchildren were asked to name the species of newly born animal that they liked the most. The results of the survey are presented in the following bar graph.

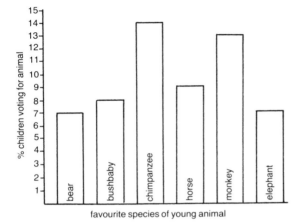

favourite species of young animal

a) By how many percent was the most popular species of animal the clear favourite over its nearest rival?
b) What total percentage of the children voted for the chimpanzee, the monkey and the bushbaby?
c) What percentage of the children voted for other animals not shown in the bar graph?
d) In reply to the question 'Which species of young animal do you dislike the most?', 540 children out of the 2000 chose the baby snake. Express this number of children as a percentage of the total.
e) 8% of the children decided that the tarantula spider was their least favourite baby animal. How many children held this view?

2 Draw up a table with the name of each of the following nine people in the extreme left hand column. Complete the table to show the five differences stated in the diagram that exist between the nine people.

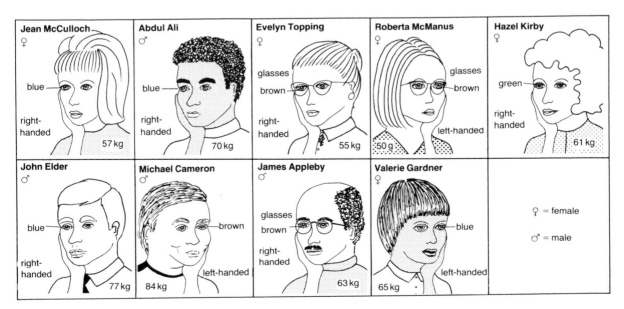

3 The following table shows the percentage distribution of blood groups in the United Kingdom.

	blood group type (%)			
	A	B	AB	O
Scottish	34	11	3	52
English	42	8	3	
Irish	26		7	32
Welsh	38	10		49

a) Present the Scottish figures only, as a bar graph.
b) Copy and complete the table.
c) Calculate the UK average percentage for each blood group and extend your table to include these figures.
d) Is blood group an example of continuous or discontinuous variation?

4 The following histogram shows the variation in mass of the individual grapes that made up a bunch.

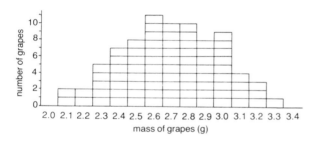

a) Describe the distribution of grape mass in terms of range and most common value for this population.
b) How many grapes were found to have the most common value?
c) What was the least common mass in this survey?
d) How many grapes had a mass of 2.8 g?
e) How many grapes were weighed?
f) What percentage of grapes had a mass of 2.5 g?
g) How many grapes weighed more than 2.7 g?

Extra Questions
⊞

h) Calculate the average mass of a grape for this population.

5 A species is a group of organisms whose members are able to interbreed and produce fertile offspring.
  The table shows the results of crosses between eight different types of animals. A shaded box means that fertile offspring resulted from the cross. A blank box means that the cross failed (i.e. sterile or no offspring produced).

animal type

	1	2	3	4	5	6	7	8
1	▨				▨			
2	▨							
3			▨	▨				
4			▨					
5	▨				▨			
6						▨		
7							▨	▨
8							▨	▨

(animal type — vertical axis)

How many different species of animal were present?

6 The lengths of 60 leaves from an apple tree are listed in the table below in random order. Divide the group up into 9 sub-sets differing from each other by 10 cm (with the first sub-set being 21–30 cm) and then present the information as a histogram.

leaf no.	length (cm)	leaf no.	length (cm)	leaf no.	length (cm)
1	48	21	90	41	92
2	31	22	97	42	63
3	52	23	66	43	69
4	61	24	59	44	69
5	57	25	56	45	60
6	77	26	64	46	57
7	86	27	62	47	43
8	81	28	75	48	67
9	103	29	83	49	70
10	72	30	99	50	80
11	80	31	33	51	84
12	68	32	45	52	47
13	65	33	28	53	79
14	58	34	42	54	76
15	45	35	56	55	95
16	39	36	61	56	58
17	62	37	64	57	55
18	61	38	77	58	74
19	76	39	89	59	54
20	78	40	81	60	78

21 What is inheritance?

1   In a genetics experiment, a pupil crossed 6 ebony-bodied male fruit flies with 1 wild type female in a culture tube. While the tube was in the incubator at 25 °C, the female laid 59 eggs. After 10 days the pupil inspected the tube and found it to contain 60 wild type flies and 6 ebony-bodied. She concluded therefore that these flies made up the F_1 generation.
a) Spot an error in the girl's experiment that accounts for the strange result.
b) Assume that you have been asked to repeat this experiment. In addition to correcting the girl's procedural error, name ONE further change that you would make to improve the experiment.

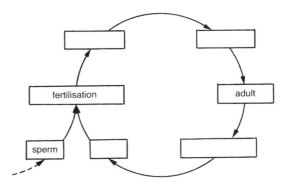

2   Read the passage and answer the questions based on it.

In 1866 Gregor Mendel, an Austrian monk, published the results and conclusions from breeding experiments on pea plants. His work is now regarded as brilliant for its time.

He understood the importance of studying only one difference between parent plants at a time. Not content with describing the offpsring produced, he also carefully counted them making it possible to express his results quantitatively (e.g. as a ratio).

Without knowing about chromosomes or genes, he was able to interpret his results and from them make up laws of genetics that still stand to this day. However during his lifetime, Mendel was not regarded as an important scientist and it was not until 1900, 16 years after his death, that his work was found by a Dutch biologist and recognised for its true worth.
a) Choose the best title for the above passage from the following:
A Sexual reproduction in pea plants
B Mendel, the father of modern day genetics
C Famous scientists of the 19th century
D Why Mendel's work remained ignored.
b) Give TWO examples of sound scientific procedure followed by Mendel in his experimental work.
c) Suggest why Mendel is regarded as 'a man ahead of his time'.

3   Copy and complete the accompanying diagram of the human life cycle, using the following terms to fill the blank boxes: egg, growth, zygote, gamete formation.

4   The following diagrams show the chromosomes present in the nuclei of various types of cells from the fruit fly, Drosophila.

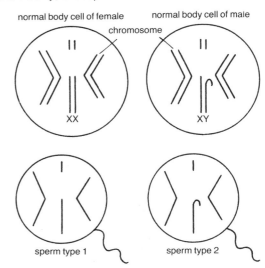

a) (i) Which of the following diagrams represents the set of chromosomes that would be found in a zygote formed as a result of sperm type 1 fertilising a normal egg?
(ii) State which sex this zygote would be.
b) (i) Which of the following diagrams represents the set of chromosomes that would be found in a zygote formed as a result of sperm type 2 fertilising a normal egg?
(ii) State which sex this zygote would be.

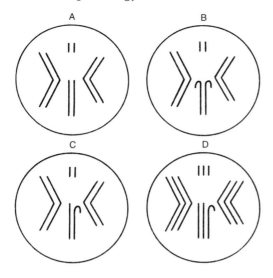

5 In pea plants the gene for height has two gene forms (alleles), tall and dwarf. The following cross was carried out:

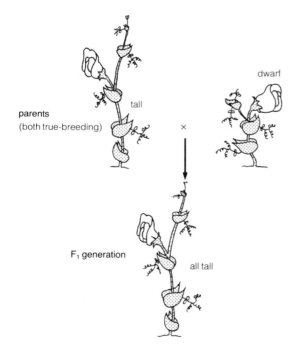

parents
(both true-breeding)

tall   ×   dwarf

F_1 generation   all tall

a) Which gene form (allele) is dominant? Explain your answer.
b) Using letters of your own choice, give the genotype(s) of the F_1 generation.

c) A further generation of pea plants was produced by allowing the F_1 generation to self-pollinate. Using your chosen symbols, copy and complete the following table to show the outcome of this cross.

	genotypes of pollen	
genotypes of ovules		

Extra Questions

d) The F_2 generation consisted of 840 plants. In theory, how many would be (i) homozygous for tallness, (ii) heterozygous for tallness, (iii) dwarf?

6 In humans, brown eye colour is dominant to blue eye colour. A brown-eyed woman marries a blue-eyed man and they have four children. One son and one daughter both have brown eyes and one son and one daughter both have blue eyes.
a) Present this information as a family tree, giving a key to all the symbols that you use.

Extra Questions

b) Using letters of your choice, label the tree with the genotypes of (i) each parent, (ii) each child.
c) Explain how you arrived at your answer to part (b) (ii).

7 The results displayed in the table opposite were obtained by first crossing a true-breeding black mouse with a brown mouse. (Black is dominant to brown.)
   The members of the F_1 generation consisted of 3 males and 5 females. These were used in turn as the parents of the F_2 generation.
a) Total up the F_2 and express it as a phenotypic ratio.
b) Suggest why it is better to consider ten F_2 litters instead of just one.

	number of black mice	number of brown mice
original parents	1	1
F_1	8	0
F_2 litter 1	5	3
litter 2	5	3
litter 3	7	0
litter 4	7	2
litter 5	5	2
litter 6	6	0
litter 7	5	4
litter 8	7	2
litter 9	7	2
litter 10	6	2

 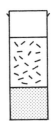

A containing 6
♂ wild type and 14
♀ dumpy-winged

B containing 17
♂ dumpy-winged
and 3 ♀ wild type

 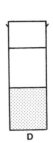

C D

8 A pupil was asked to design an experiment to show that crossing male wild type fruit flies with female dumpy-winged flies is genetically the same as crossing male dumpy-winged flies with female wild type flies. He was given the apparatus shown below (where all the flies are true-breeding).

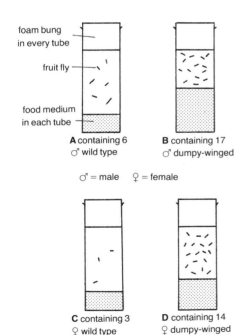

foam bung in every tube

fruit fly

food medium in each tube

A containing 6
♂ wild type

B containing 17
♂ dumpy-winged

♂ = male   ♀ = female

C containing 3
♀ wild type

D containing 14
♀ dumpy-winged

The following diagram shows what he proposed to set up as his test.

State TWO ways in which his experimental set-up would have to be altered to make it fair.

9 In each of the following crosses, the original parents were true-breeding and the members of each F_1 generation were crossed with one another.

Example 1:
Mouse fur colour where black is dominant to brown

	number of black	number of brown
parents	1	1
F_1	9	0
F_2	60	20

Example 2:
Guinea-pig hair type where straight is dominant to wavy

	number of straight	number of wavy
parents	1	1
F_1	7	0
F_2	29	11

Example 3:
Hamster fur colour where golden is dominant to white

	number of golden	number of white
parents	1	1
F_1	8	0
F_2	45	16

	number of brown	number of grey
parents	1	1
F_1	6	
F_2		8

a) What generalisation about the phenotypic ratio of the F_2 generation can be drawn from this information?

b) In gerbils brown coat colour is dominant to grey. Copy and complete the following table where a true-breeding brown gerbil is crossed with a grey mate and their offspring are then crossed with one another, by inserting the expected numbers.

10 In pea plants, the allele for flower colour (**C**) is dominant to the allele for lack of flower colour (**c**).
   A plant homozygous for flower colour was crossed with a plant bearing colourless flowers. The F_1 plants were then self-pollinated.
   Which of the following correctly represents the ratio of genotypes expected in the F_2 generation?

A   all **Cc**   B   1**CC**:1**Cc**   C   3**CC**:1**cc**
D   1**CC**:2**Cc**:1**cc**

22 Genetics and society

1 Read the passage and answer the questions based on it.

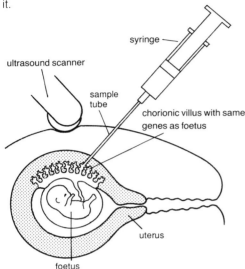

ultrasound scanner

syringe

sample tube

chorionic villus with same genes as foetus

uterus

foetus

A new technique for diagnosing genetic diseases in unborn children is known as chorionic villus (CV) sampling. A tiny sample of tissue from the placenta is removed from the mother's womb using a special syringe guided by ultrasound scanners as shown in the accompanying diagram.

The cells (genetically identical to the foetus) are inspected in the laboratory to see if their genetic material has unusual characteristics that indicate that the baby will suffer from an incurable disease.

If this proves to be the case, the parents can decide to have the pregnancy terminated. This technique can be done as early as 8 weeks into the pregnancy when an abortion may be less distressful to the mother than at 18–20 weeks (which is when the alternative test, amniocentesis, is normally done). However, initial studies have shown that CV sampling is followed by a miscarriage in about 3 out of every 100 of those mothers whose babies are found to be normal.

a) By what means can a sample of tissue, genetically identical to an unborn baby, be obtained without touching the baby?
b) Why is this tissue inspected in the laboratory?
c) Suggest another name for the 'genetic material' referred to in lines 8 to 9.

d) Give ONE way in which the above technique is preferable to amniocentesis.
e) State a disadvantage of CV testing.

2 Read the following passage and answer the questions based on it.

In 1791, a sheep farmer discovered a completely new variety amongst his flock. A male lamb had been born with short legs. We now know that this new form of leg length had arisen as a result of the lamb inheriting a changed gene which neither parent possessed.

The farmer decided that he would like a whole flock of short-legged sheep since the low fences needed to keep them in the fields would save him money on materials. He therefore crossed the unusual ram with a normal ewe as shown in the diagram.

● = gene form for short legs
○ = gene form for long legs

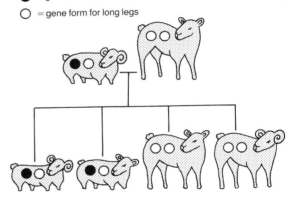

He then crossed only the short-legged animals with one another for several generations and eventually produced a flock of short-legged sheep.

a) Which words in the passage mean a 'mutation'?
b) Is the gene form (allele) for short legs dominant or recessive? Explain your answer with reference to the diagram.
c) Using the circular symbols, make a keyed diagram to show the outcome on average of crossing the two short-legged animals shown in the diagram. State the phenotype of each offspring produced.

Extra Questions ⊞

d) (i) Is short-leggedness in sheep an example of advantageous or disadvantageous variation to the farmer? Explain your answer.
(ii) Would short-leggedness in sheep be advantageous or disadvantageous to wild sheep before domestication by man? Explain your answer.
e) Which sentence in the passage refers to an example of selective breeding over a long period of time?

3 Cystic fibrosis is an inherited disorder of the human body. It affects mucus production, causing blockage of tiny air passages in the lungs. It is due to a recessive gene form.
   Consider the following family trees:

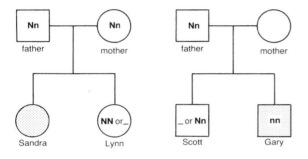

☐ = normal male

◯ = normal female

▨ = male affected by cystic fibrosis

◌ = female affected by cystic fibrosis

a) Copy and complete the diagram by inserting the missing genotypes.

Lynn and Scott are intending to marry so they seek the advice of a genetic counsellor. The counsellor studies their family trees and figures out that if the couple marry then this would result in one of the following crosses:

(i) NN × NN or (ii) NN × Nn or (iii) Nn × Nn

b) Which TWO crosses involve no risk of producing affected children?
c) In the remaining cross, what is the chance of each child being affected?
d) Show in diagrammatic form how you arrived at your answer to (c).

4 The graph shown below refers to an investigation into the effect of increasing radiation on the percentage number of X chromosomes showing a lethal (deadly) mutation.
   The animal used was the fruit fly (*Drosophila melanogaster*).
   The results are plotted as points with the best straight line drawn through them.

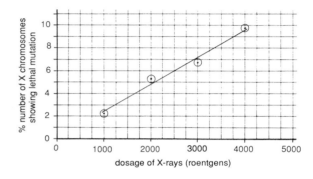

a) What was the one variable factor controlled by the experimenter in this investigation?
b) How many conditions of this factor were used?
c) Name TWO other factors that would have to be kept constant to make the experiment fair.
d) What conclusion can be drawn from the above results?
e) Suggest a suitable control for the experiment.

Section 7 Biotechnology

23 Living factories

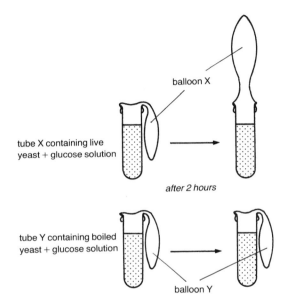

balloon X

tube X containing live
yeast + glucose solution

after 2 hours

tube Y containing boiled
yeast + glucose solution

balloon Y

1 The experiment shown in the accompanying
diagram was set up to investigate the effect of live
yeast on glucose solution. It was left in an incubator
at 30 °C for two hours.
a) Explain why balloon X has inflated but balloon Y
has failed to do so.
b) Predict the effect of repeating the experiment at
(i) 10 °C and (ii) 70 °C on the time required to inflate
balloon X to the size shown in the diagram. Explain
your answer in each case.
c) With reference to the glucose solution, identify a
feature not shown in the diagram that must be kept
constant in tubes X and Y for a valid comparison of
the results to be made.

2 The following table refers to six types of bacteria.

scientific name of bacterium	shape	motile?	colour after gram stain	spore-forming?
Lactobacillus	rod	−	purple	−
Clostridium	rod	+	purple	+
Streptococcus	sphere	−	purple	−
Salmonella	rod	+	pink	−
Staphylococcus	sphere	−	purple	−
Bacillus	rod	+	purple	+

(+ = yes, − = no)

a) Which type of bacterium fails to become purple
after gram staining?
b) Consider the sphere-shaped bacterial types in
the table. Apart from shape and becoming purple
after gram staining, what TWO other features do
they have in common?
c) Consider the spore-forming types of bacteria
and identify THREE other features that they have in
common.
d) Which type of bacterium is rod-shaped, purple
after staining but fails to make spores?
e) How many types of bacteria are motile and
spore-forming?
f) Identify TWO different pairs of bacterial types that
each have all four features in common.

3 The experiment shown in the diagram overleaf was
set up to investigate whether live yeast produces
heat when respiring anaerobically (i.e. in the total
absence of oxygen). State TWO ways in whch the
experiment will have to be altered in order to make it
a fair test.

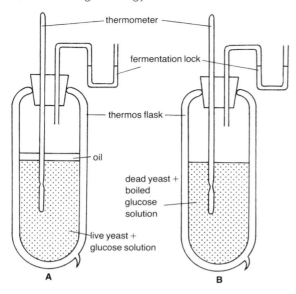

A   B

time (hours)	rate of CO_2 production (cm³/hour)
0	0
10	30
20	60
30	90
40	110
50	116
60	116

a) Using graph paper similar to that shown, and the axes (which have been partly completed for you), present the above results as a line graph.

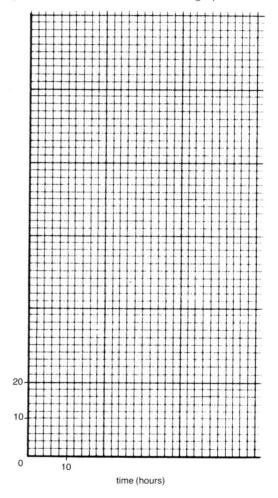

4 Sugar is broken down into lactic acid by a type of bacterium called *Lactobacillus*. Calcium carbonate is a white insoluble solid which reacts with acid to form a clear colourless solution.

A plate of agar containing sugar and calcium carbonate (which gives the agar a cloudy appearance) was prepared and then four different types of bacteria (A, B, C and D) were added to it as shown in the following diagram

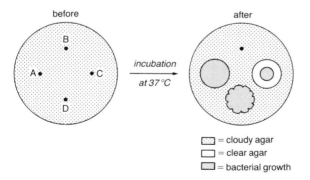

□ = cloudy agar
□ = clear agar
▨ = bacterial growth

a) Identify *Lactobacillus*.
b) Give a reason for your choice.

5 The rate of carbon dioxide production from a yeast culture was measured at regular intervals over a period of 60 hours and the results tabulated as follows.

b) How long did it take until CO_2 was being produced at a rate of 84 cm³/hour?
c) State the volume of CO_2 being produced per hour at (i) 14 hours after the start, (ii) 56 hours after the start.

6 Natural yoghurt contains live bacteria which convert sterile milk into yoghurt. A boy was asked to set up an experiment to find out the best (optimum) temperature for yoghurt formation. He rinsed three glass beakers with hot water and dried them on clean paper towels. He then used them to set up the following experiment.

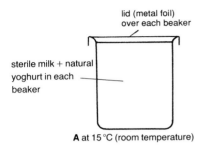

lid (metal foil)
over each beaker

sterile milk + natural yoghurt in each beaker

A at 15 °C (room temperature)

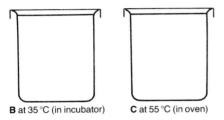

B at 35 °C (in incubator)   **C** at 55 °C (in oven)

Results (after 6 hours)

	texture of 'yoghurt'	smell of 'yoghurt'
A	watery	faintly unpleasant
B	creamy	unpleasant
C	watery	no smell

a) Spot the important error in the boy's procedure that probably led to the unpleasant smell developing.

Extra Question

The error was corrected in a second attempt at the experiment and the smell did not develop. This time the results were as follows.

	texture of 'yoghurt'
A	watery
B	creamy
C	watery

The boy concluded that 35 °C is the best temperature for yoghurt formation.
b) Suggest how his experiment could be further improved to find out a more accurate estimate of the optimum temperature.

24 Problems and profit with waste

1 A pupil was asked to find out if sour milk contains more bacteria than fresh milk. The following series of diagrams shows the procedure that he followed.

Spot TWO mistakes that the boy made and then copy and complete the table at the foot of the page.

1 He collected the following apparatus

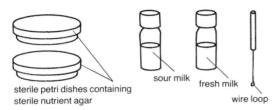

sterile petri dishes containing
sterile nutrient agar

sour milk   fresh milk

wire loop

2 He held the wire loop in a Bunsen flame until it was red hot and then cooled it by waving it in the air

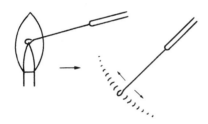

3 He collected a small drop of sour milk on the loop

sour milk

4 He spread the sour milk over the nutrient agar in one of the plates

5 He then used the wire loop to take a sample of fresh milk which he spread in the same way onto the second plate of agar.

6 He sealed the dishes with sticky tape and put them in a warm incubator

2 The rate at which an organism increases in dry weight is often taken as an indication of growth. The maximum percentage dry weight increase shown by different organisms (all grown under ideal conditions for 24 hours) varies enormously. Large multicellular animals grow much more slowly than tiny unicellular organisms. The foetus of a cow, for example, only increases by 1 per cent of its dry weight over 24 hours whereas yeast cells increase by 1400 per cent and bacteria by 4750 per cent of their dry weight over the same period. When compared with microbes, multicellular plants similarly show a modest percentage dry weight increase. A Scots pine tree only gains 5 per cent and even a fast growing plant such as sunflower 29 per cent in 24 hours.

Present the information contained in the above passage in a table so that the data can be easily compared.

error or omission in procedure	correct procedure	reason for following correct procedure

3 Read the following passage and answer the questions.

Single-celled protein (SCP) is produced by certain strains of unicellular microbes cultured on petroleum hydrocarbons. Vast amounts of this protein, which is used for animal feed, are produced quickly, since single-celled micro-organisms grow at a very rapid rate. The protein content of the product is high and its production is independent of climate. However, SCP contains a high concentration of nucleic acid. If eaten by humans this would be converted to uric acid which would gather as crystals in the joints causing a painful condition called gout. In farm animals, uric acid is converted by an enzyme into a soluble waste which is later excreted.

Mycoprotein is produced by a fungus called *Fusarium* growing on glucose syrup. Although able to double its weight every five hours, this multicellular fungus grows more slowly than single-celled microbes. Its product, mycoprotein, does not contain excessive amounts of nucleic acid. Unlike unicellular micro-organisms, *Fusarium* has a thread-like structure resembling meat fibres in size and strength. It can therefore be spun into a meat-like substance by interlaying the fibres with suitable colouring and flavouring. Food technologists have already used mycoprotein to produce meatless 'burgers' and 'sausages' that contain 44 per cent protein and do not shrink on cooking. By varying the flavouring and other additives, they are now producing realistic 'chicken and ham paté', 'game pie' and even 'chocolate biscuits'. The manufacturers, concerned that some consumers might be put off these foods by the thought of eating fungus, are quick to point out that mushrooms, a traditional and widely accepted food, are also fungi.

a) Apart from high protein content, what two advantages are gained by producing SCP for animal feed rather than using a plant crop such as turnips?

b) Explain why SCP can be fed to farm animals yet is unsuitable for human consumption.

c) Unlike SCP, mycoprotein is safe for humans to eat. Explain why.

d) (i) Describe the texture of the fungus *Fusarium*.
(ii) Of what advantage is this texture to food technologists?

e) (i) Why might foodstuffs made of mycoprotein meet with consumer resistance?
(ii) What defence to such criticism is offered by the manufacturers?

4 The following experiment was set up to investigate the production of methane by microbes respiring in the absence of oxygen. It has been running for two weeks.

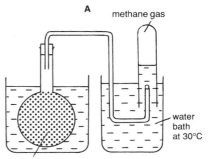

A

methane gas

water bath at 30°C

fresh cow dung + boiled and cooled tap water

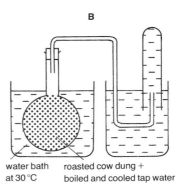

B

water bath at 30°C

roasted cow dung + boiled and cooled tap water

a) At the start, test tube A was full of water but now it contains some methane gas. Account for the fact that no methane is present in tube B.

b) Give TWO possible reasons for boiling the tap water before use in the above experiment.

5 When grown in a bioreactor at 30°C, the mycoprotein fungus doubles its weight every 5 hours as shown in the following partly completed table of results.

time (hours)	mass of fungus (g)
0	100
5	
10	400
	1600

a) Copy and complete the table.

b) Draw a line graph of this information.

c) Predict what would happen to the growth rate of the fungus at 10 °C.

Extra Question ⊞

d) On your graph paper, draw a further line that could represent the growth rate of the fungus at 10 °C.

6 The graph below shows the number of bacteria growing in nutrient broth kept at constant optimum temperature over a period of 30 hours.

a) How many bacteria were present per mm^3 at 5 hours?
b) From this point in time onwards, how many more hours passed before the bacteria had doubled in number?

Extra Questions ⊞

c) During which of the following periods of time was growth rate greatest?

A 0–5 hours   B 5–10 hours   C 10–15 hours   D 15–20 hours
d) For how many hours was the number of bacteria found to be above 4000 per mm^3?
e) Give TWO possible reasons why the number of bacteria began to decrease after 23 hours.

7 An equal volume of fresh, untreated milk was placed in each of four sterile screw-top bottles (W, X, Y and Z). The bottles and their contents were then treated as shown opposite:

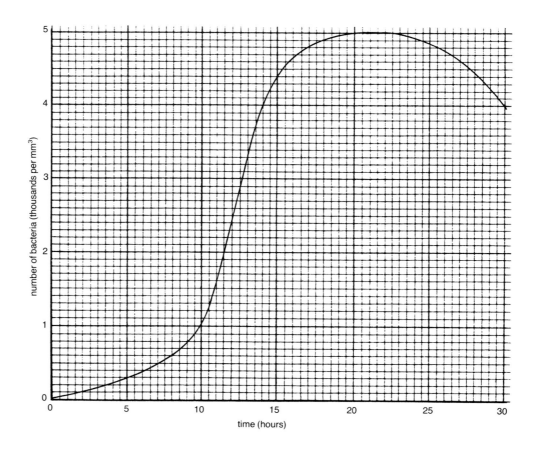

	treatment		
bottle	heated in pressure cooker at 120 °C for several minutes	pasteurised at 73 °C for 15 seconds	place where bottle was then kept for one week
W	no	yes	warm room
X	yes	no	fridge
Y	no	no	warm room
Z	no	yes	fridge

a) The milk failed to turn sour in one of the bottles. Identify it and explain your choice of answer.
b) Predict the order in which the milk in the other bottles turned sour.

8 A type of bacterium called *Escherichia coli* normally lives in the human large intestine. When special culture medium (McConkey's broth) is inoculated with *Escherichia coli* and incubated, the broth changes in appearance from clear red to cloudy yellow.

In the experiment shown in the diagram, samples of water from three rivers X, Y and Z were investigated.,

a) Construct a hypothesis (concerning the type of pollution affecting river Y) to account for the results obtained.
b) Suggest why the test tube is floating in the sample from river Y after incubation.

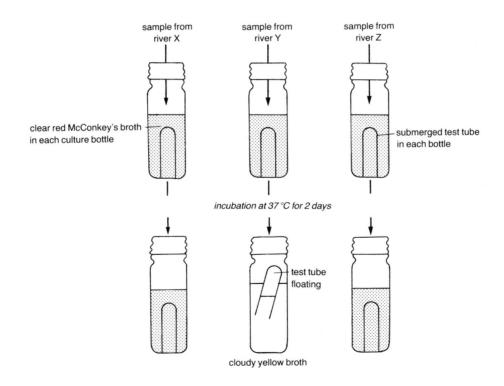

sample from river X   sample from river Y   sample from river Z

clear red McConkey's broth in each culture bottle

submerged test tube in each bottle

incubation at 37 °C for 2 days

test tube floating

cloudy yellow broth

25 Reprogramming microbes

1 Study the following shopping list and answer the questions that follow.

quantity	item	price (£)	biotechnological (√) non-biotechnological (×)
800 g	bread	0.50	√
0.5 kg	cheese	1.50	√
1 kg	apples	1.00	×
3.5 kg	soap powder	3.00	√
750 g	yoghurt	1.00	√
3 kg	potatoes	1.00	×
1 litre	fruit juice	0.50	√
2 dozen	eggs	2.50	×
750 cm³	wine	3.00	√
400 g	barbecue sauce	1.00	√
4	grapefruit	1.00	×
250 g	butter	0.50	√
300 cm³	vinegar	0.50	√
1 kg	meat	4.50	×
3 litres	beer	3.50	√

a) How many biotechnological products are on the shopping list?

b) Express the number of biotechnological to non-biotechnological products as a ratio.

c) What percentage of the total amount of cash was spent on biotechnological products?

2 The experiment below was set up to investigate the effect of two antibiotics (penicillin and streptomycin) on the growth of two species of bacteria.

State TWO ways in which the experiment must be altered in order to make it a fair test.

3 Cells from one species of bacteria were spread over the surface of sterile nutrient agar in a petri dish. A multidisc with a different antibiotic at the end of each of its six arms was then placed on top of the bacteria.

The following diagram shows the result of the experiment after 48 hours in a warm incubator.

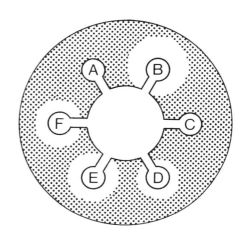

A–F = 6 different antibiotics

▓ = zone of bacterial growth

☐ = zone of no bacterial growth

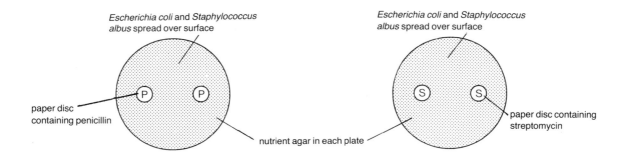

Escherichia coli and Staphylococcus albus spread over surface

paper disc containing penicillin

Escherichia coli and Staphylococcus albus spread over surface

paper disc containing streptomycin

nutrient agar in each plate

a) (i) Which antibiotics were able to prevent growth of the bacterium?
(ii) To which of these was the bacterium most sensitive?
(iii) Explain your answer to part (ii).
b) To how many of the antibiotics was the bacterium resistant?

4 Each of the dishes in the experiment below was streaked with five different types of bacteria (a, b, c, d and e). A strip containing one of four different antibiotics (P, Q, R and S) was then placed across each dish.

The diagram shows the results after incubation for 3 days at 37 °C.
a) Which antibiotic was least effective at preventing bacterial growth?
b) Which type of bacterium was sensitive to all four antibiotics?
c) Which type of bacterium was most resistant to the antibiotics?
d) Which types of bacteria were sensitive to two antibiotics and resistant to two antibiotics?

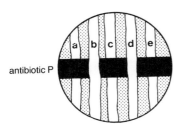

antibiotic P

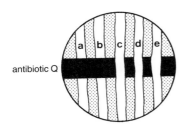

antibiotic Q

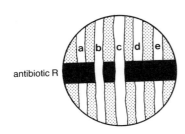

antibiotic R

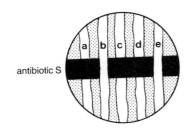

antibiotic S

5 Read the passage and answer the questions based on it.

Lactose is a sugar present in whey (the unwanted remains of milk after the curds have been used for cheese-making). It is composed of two simpler sugars. Brewer's yeast (*Saccharomyces cerevisiae*) is unable to ferment it because the fungus cannot take lactose in through its cell membrane. Even if lactose could enter a yeast cell, the organism lacks the enzyme needed to break the bond holding together the components of lactose – one molecule of glucose and one of galactose.

However, genetic engineers have now solved these problems by transferring pieces of chromosome from a rare yeast (*Kluyveromyces lactis*) into brewer's yeast. The new fungal strain formed possesses the genes required for making both lactose permease (the enzyme that must be present for lactose to enter the cell) and lactase (the enzyme which digests lactose to glucose and galactose). As a result normal fermentation can now proceed.

a) Describe the structure of a molecule of lactose.
b) Give TWO reasons why *Saccharomyces cerevisiae* is unable to use lactose as a foodstuff.
c) With reference only to the information in the passage, describe what is meant by the term genetic engineering.
d) Give TWO reasons why the new strain of yeast formed by genetic engineering is able to make use of lactose.

Extra Question

e) Which of the following chemical reactions is this new microbial strain able to bring about?

A lactase → alcohol   B alcohol → galactose
C galactose → glucose   D glucose → alcohol

6 The four experiments in the following diagram were
set up to investigate the action of a new 'biological'
soap powder called BIOX. Study them carefully and
answer the questions that follow.

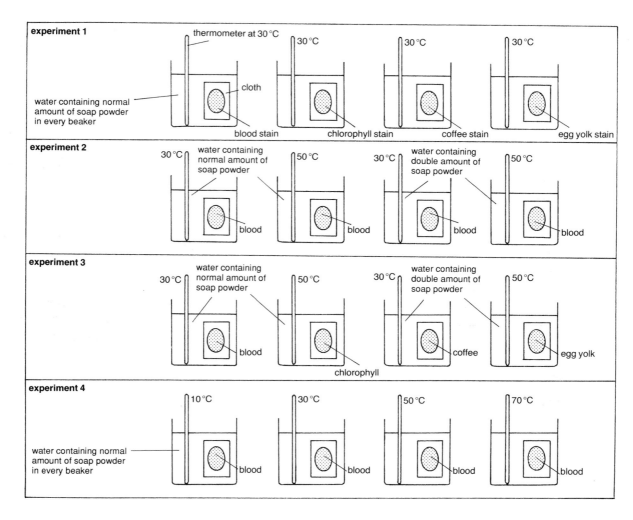

a) Which experiment is testing the effect of four
different temperatures on the action of BIOX?
b) In which experiment is the type of stain the one
variable factor being investigated?

Extra Questions

c) Which experiment really consists of two
experiments being done at the same time?
d) Which experiment is invalid? Explain why.